Cosmic Life-Force

Cosmic Life-Force

Fred Hoyle and
Chandra Wickramasinghe

Paragon House
New York

First American edition, 1990

Published in the United States by
Paragon House
90 Fifth Avenue
New York, NY 10011

Originally published in the United Kingdom in 1988 by
J.M. Dent & Sons, Ltd.

Library of Congress Cataloging-in-Publication Data

Hoyle, Fred, Sir.
    Cosmic life-force / Fred Hoyle and Chandra Wickramasinghe. — 1st
  American ed.
        p.   cm.
    Reprint. Originally published: London : J.M. Dent, 1988.
    Includes bibliographical references.
    ISBN 1-55778-226-0
    1. Life—Origin.   2. Life on other planets.   I. Wickramasinghe,
  N. C. (Nalin Chandra), 1939-   . II. Title.
  QH325.H675   1990
  577—dc20   89-38448
                                                            CIP

Manufactured in the United States of America

Contents

Illustrations

Introduction

It is over a decade since we launched our original ideas on the cosmic beginnings of life. Our first book in the present series, *Lifecloud*, dealt mainly with the origins of the chemical building blocks of life – e.g. molecules like cellulose – which we considered to be widespread in our galaxy and beyond. Our seemingly heretical point of view was resisted, but only somewhat mildly, compared with what was to follow in later years. The competing theory at the time was that the organic molecules that were required for an origin of life were synthesised *in situ* on the Earth from a mixture of inorganic materials such as water, methane and ammonia exposed to energetic events such as lightning. Already in 1978 we found ourselves being slated by reviewers who had no hesitation in advocating the use of Ockham's Razor to eliminate what they considered to be a redundant and unprofitable hypothesis.

The tide of events was soon to prove our opponents wrong, however. Not only was the widespread occurrence of organic molecules in our galaxy established beyond a shadow of doubt, but such molecules – some of them exceedingly complex – were identified in extraterrestrial objects nearby. Complex organic molecules were found first of all in carbonaceous meteorites, then in extraterrestrial particles collected in the upper atmosphere, and later in comets. Events of 1986, including the direct probing of material from Comet Halley, culminated in a general acceptance, albeit reluctantly, of 'organic' comets. The ideas of *Lifecloud* which were thought highly controversial in 1977 slipped quietly and without ceremony into the realms of ortho- dox science.

The further development of our thinking in the years after 1977 led to certain ramifications of these ideas that came into direct head-on collision with long-established cultural values. The publication of *Diseases from Space* triggered a torrent of hostility for the reason that one is apparently not permitted on any account whatsoever to revive ideas that have once been discarded as medieval superstition. It was (and still is) our belief that the relevant factual evidence points strongly to an extrater- restrial origin of viruses and bacteria, and that on occasion the incidence of such microorganisms from space could explain the phenomenon of epidemic disease. Doctors and epidemiologists in the last century who described the spread of influenza as a 'miasma' descending over the land were not far off the mark, we think.

In *Evolution from Space* we again touched a raw nerve in our critics because our ideas could be interpreted with some licence as advocating a return to pre-Darwinian thinking. The creative driving force of evolution, we had argued, was derived not from mistakes in copying old genetic information as is currently thought, but from the addition of new genes in the form of vi- ruses and bacteria from space. Although it is the case that neither Charles Darwin nor Alfred Russel Wallace actually specified the ultimate source of innovative mutations, later workers assumed without proof that such innovations must all be generated intern- ally within the framework of terrestrial biology. Our challenge of this well-entrenched dogma was seen by many as an abroga- tion of a position that was reached after a long and strenuous struggle in the closing decades of the last century. In no way could they contemplate throwing aside a victory that was so hard won. There was also a danger that was perceived in relation to events in the present day. Any concession to the cosmic life theory tends to be resisted nowadays for fear that it could throw open the floodgates to creationist passion. Creationism of the kind espoused by fundamentalist groups in the United States is not a doctrine to which we ourselves can even remotely sub- scribe. Yet we do not think that the pursuit of science in an objective way can or should be sacrificed for fear that its discoveries may give succour to objectionable social groups.

Notwithstanding the many sociological obstacles in our way, we have pursued our ideas logically and relentlessly, leading eventually to the conclusions that are set out in the present book. Data gathered from a variety of scientific disciplines from the time of *Lifecloud* to *Cosmic Life-Force* have served as a basis for verifying our original thesis as well as for extending these ideas in directions that could not have been foreseen in 1977. We offer the reader an account of what might be described as our final synthesis, connecting the evolution of life and intelligence on the Earth with processes that are cosmic on the grandest possible scale.

ONE

The terrestrial abode

In this book we shall argue that life is a phenomenon that must involve the entire Universe. The living process as we know it cannot be confined to events that occur on an individual planet like the Earth, and attempts to constrain it in such a manner lead to serious conflicts with experimental data.

What is life? This is not an easy question to answer, and one could wonder whether there will ever be a precise answer. Even the very simplest life-forms are comprised of an incredibly complex, well-ordered set of chemicals that are constantly reacting with each other, seeking amongst other things to replicate their entire structures. Chemists and biochemists can tell us a great deal about the nature of these chemicals, but it is almost impossible with present-day methods to follow all the complex changes that take place as cells perform their various functions. For a definition of life we shall adopt in this book the pragmatic one of recognising what we know to be unmistakably living, what everyone will accept as life without dispute.

Life has properties that are unmistakable at the level of individual cells, and also cooperative behavioural properties that can be identified with organisms, groups of organisms or even entire ecologies. Amongst the human species there are many properties and traits that are common to lower animals, whilst others are so different as to warrant special attention. Our love of art, music and abstract contemplation fall into this latter category.

The modern conservationist, for instance, is attempting to

impose aesthetic value judgements on the world of life. The motivations of such a person are part animal, part super-animal, as we can see. Ardent conservationists who seek to preserve a pristine countryside or a scenic coastline from the ravages of present-day human society are fighting a losing battle. They are prominent on public platforms the world over and their pronouncements are frequently displayed on the front pages of our daily newspapers, but it takes scarcely a cursory glance to show that they are not really making much progress. A silent all-pervading force is at work paying no heed whatever to their vocal protestations. The situation becomes amply clear as one flies in an airplane over the built-up cities of the so-called civilised world, as for instance in the United States of America. Cities are in a state of rapid expansion, encroaching more and more upon forests, greenbelt and natural landscapes of various kinds, slowly but surely replacing them with a maze of concrete jungle and all the attendant social problems that such conurbations imply. There seems at the moment no limit to this inexorable process except in respect of the minimum needs of human beings to maintain agriculture – land for growing crops and grazing livestock all cramped into minimal sized compartments by modern techniques of intensive farming.

The driving force behind the conservationist is apparently some inner need to live in harmony with Nature and to preserve and protect the great multitude of species of plants and animals besides Man. Live and let live is his golden motto, not only for what is here and now but for all future generations as well. Antithetical to conservationism is a powerful expansionist force that seeks to separate Man from other life-forms and urges him to assert his absolute ascendency, destroying the rest in the process.

This seemingly arrogant and selfish attitude that controls modern society is ironically an outgrowth of the process of human civilisation, a product of the increasingly complex levels of social organisation that have come to exist. Reprehensible though it may be, it is a derivative of a cosmic life-force that operates even at lower levels of life but to a somewhat lesser degree. This 'force' seeks at all costs to remould and modify the environment to suit itself by an imperative need to increase to the greatest possible extent the population of a particular type of living system — a genotype. This process operates unremittingly where it can, so the conservationist can have little hope of success in the present state of the world.

Some four billion years ago the Earth was a barren lifeless planet, a prospect not unlike a lunar landscape in its broader features. The first step towards remoulding our planet occurred as a result of the arrival of comets that carried volatile materials from the outer regions of the solar system. The water that came from comets went to form the Earth's oceans. The evaporation of water from these oceans, followed by the break-up of water molecules by sunlight, provided just the right conditions for the development of a terrestrial atmosphere that eventually became conducive to primitive life. At the same time there started the evaporation–precipitation water cycle and the laying down of soils through the processes of weathering of the Earth's crust and the accumulation of sediments.

The fact that the oldest sedimentary rocks on the Earth, which were laid down some 3500 to 3800 million years ago, show the presence of microbial fossils is, in our view, a clear indication that life was added to the Earth from outside. There was little or no time available for any 'primordial soup' to have brewed on the Earth prior to the time when these sediments were laid down. Microorganisms appear to have been incident on the Earth from the very beginning of its existence as a planet, but their taking root was contingent on the initial remoulding that has been already mentioned. The atmosphere was clearly needed. Life shows up in the rocks close to the first moment when it was possible to survive. We know from studies of lunar rocks that from the time of the formation of the Earth, some 4600 million years ago, to about 3800 million years ago, both the Moon and the Earth must have suffered such severe bombardment by meteorites that a quiescent terrestrial crust or an atmosphere would not have been possible. In later chapters we shall show that the presence of an atmosphere to cushion the landings of microorganisms is a primary requirement for the survival of cosmic life.

In view of the likelihood that the original atmosphere was lacking in ozone, the gas that shields the solar ultraviolet rays, it seems highly probable that the first major explosion of life occurred in the oceans. Below several centimetres of ocean the condition for ultraviolet shielding would have been automatically satisfied. The development subsequently of an atmosphere which included ozone permitted the expansion of life from ocean to land.

The primary aim of life would seem to be directed at converting an aqueous inorganic environment to life of a similar kind. The simplest functional unit of life might be regarded as the cell which possesses a porous organic membrane surrounding the genetic material, proteins, sugars and the rest. To make organic matter from water, carbon dioxide and inorganic salts one requires energy, and this energy has to be ultimately derived from sunlight. The so-called photosynthetic organisms – plants, algae and certain types of bacteria – utilise the energy of sunlight by means of the complex organic molecule known as chlorophyll, the green substance of plants. The first step in photosynthesis is a reaction of the type:

$$\text{water} + \text{carbon dioxide} + \text{chlorophyll} + \text{sunlight}$$
$$\rightarrow \text{glucose} + \text{oxygen} + \text{chlorophyll}$$

In this reaction the chlorophyll molecule acts as a catalyst by providing in essence a storage battery for solar energy and releasing this energy so as to bring about a chemical transformation that is vital to all life. Sugar molecules can be polymerised into starch, cellulose and other polysaccharides. Energy stored in the chemical bonds of polysaccharides or of sugars is then used to drive the complex machinery of reactions in the cell. To realise this stored energy a cell often uses the intermediary 'complex molecule' known as ATP (adenosine triphosphate).

With the beginnings of photosynthesis established in the oceans by plankton and photosynthetic bacteria, the way lay open for further developments in the elaboration of life. Photosynthetic microorganisms use the energy of sunlight ultimately to produce complex organic material on which other types of organisms can thrive. Photosynthesis also allows atmospheric carbon dioxide to be reconverted to oxygen, the gas that is required for respiration by all higher animals.

For an organism to stay alive it is necessary for it to have access, either directly or indirectly, to the energy of sunlight. The basic strategy of life is to devise ways and means for this to be done. Every single species can be looked upon as an attempt to harness what energy is around and available for it to take. The present-day Earth is inhabited by many millions of different species dependent upon one another, all scrambling to grasp what they can of the energy of sunlight that falls on its surface. The evolutionary history of this entire ensemble of life can be seen as a prudent and economical harnessing of energy and of natural resources to suit the various life-forms that came into existence at different times.

Before higher plant and animal life-forms became developed it would have been important to establish organisms such as lichens which grow on bare rock, and which are capable of taking up even small quantities of moisture in the air. Lichens produce acids which can break down the mineral structures within rocks into small particles, whilst also using sunlight via chlorophyll to manufacture organic material. When lichens die, the decomposed organic matter, together with the newly generated soil particles, offers considerable scope for more complex plant and animal life to develop. Thus higher plants such as grasses and ferns develop, eventually paving the way for the appearance of oaks, pines, maples and beeches. An enormous variety of habitats for insects and animal life are also generated in the process. Provided the genetic raw material is forthcoming one could envisage a situation where an ecology comprised of microorganisms, plants and animals develops so as to cover the entire surface of the planet. Evolutionary changes along these lines have actually occurred over long geological timescales.

In the present day both the chemical composition and detailed physical state of a layer of our planet which extends about 5 miles above the ocean surface and perhaps 5 miles below it are sensitively controlled by the forces of life. Almost every conceivable 'niche' for life has become colonised with an almost incredible range of plant, animal and microbial life. The power of life to alter the state of the Earth, even over relatively short timescales, is manifestly clear. Any gross disturbance to an established local ecology that is caused by either natural or man-made events results in rapid readjustments taking place. During the early

history of the United States of America much of the virgin forest
of New England was cleared for use as farmland, and with later
migrations westwards vast tracts of such reclaimed farmland
were abandoned. Natural recolonisation then began in a dramatic way. Hardy grasses and weeds first took over, then shrubs,
followed by junipers, poplars and the like, finally to be colonised
by the dominant beech and maple.

No matter where one turns to, in whatever corner of the Earth,
interdependent species of plants, animals and microorganisms
are in a constant state of flux. They are continually rearranging
their relationships one to another and to their environments.
Seemingly tranquil inland lakes are being continually worked
over, with ecological changes being wrought over relatively short
timescales. Consider, for instance, a lake initially surrounded
by a sturdy beech and maple forest. Photosynthetic algae and
diatoms are of course the mainstay of such a lake community;
they are the primary producers of energy. They are endowed
with the biochemical apparatus for tapping the energy of visible
sunlight and storing this in the form of organic molecules, sugars
and carbohydrates on which the so-called non-photosynthetic
life-forms can feed. They are at the very top of the ecological
energy ladder. The immediate consumers of this energy are the
ciliates and other single-celled protozoa, then worms and fish.
As time goes on, humus from decaying plants builds up around
the edges of the lake, leading to an outer rim becoming colonised
by mosses and lowly shrubs. Gradually the lake fills in from
the outside, becoming colonised by plants that eventually give
way to the dominance of the beech and maple forest throughout.

In the animal world as in the plant world, the big and strong
eventually dominate the scene, shouldering away unwanted
competition in the process. Man's own dominance on this planet
and his role in actually changing the gross physical structure
of the Earth has developed in slow stages from his first appearance some 2 million years ago. Like the ciliates and worms and
other higher forms of animal life, Man is a consumer of energy,
not a primary energy producer. Our initial ascendency over our
competing animal consumers had of necessity to depend upon
our greater skills in harnessing the sources of stored energy in
other animals and plants.

We are parasites, pure and simple, leading as it were a predatory existence on living forms that have more direct access to
the bounty of solar energy. We lack the ability to sustain ourselves without the assistance of lower animal and plant life.
Today we are perched on top of the pyramid of terrestrial life,
occupying the place of the grand predator, the unquestioned
master species of our planet. The destiny of our planet is ostensi-
bly in our hands.

For life-forms such as ourselves survival is a mad scramble
for what stored energy can be got. The more successful we are
in this act of piracy, the more dominant we become. For over
99 per cent of the time that human beings have existed, survival
of our species has not been significantly different from that of
lower animals with whom we competed. Man hunted and preyed
on other wild animals and ate whatever wild plant foods there
were to be gathered. Groups of humans at this stage tended
to live in nomadic tribes no more than a few thousand strong.
They probably lived within temporary shelters which were
moved around with the seasons in search of locations where
food was at hand. From the earliest archaeological remains it
is clear that the tools that were invented to help in the hunt
were not lacking in ingenuity or craftsmanship. Nor did they
lack artistic merit, with intricate engravings that could hardly
have served any practical purpose. From the oldest extant cave
paintings it is clear that human beings were quick off the mark
to develop some sort of religious instinct. Man looked upward
to the Heavens to seek a Creator. He would seem to have stood
apart from other creatures in addressing deep questions relating
to his own origins and to his ultimate fate. He developed magico-
religious beliefs and practices which show up in the form of
deliberate and intricate burial rites in diverse Stone Age cultures
well over 10,000 years ago. Despite all these advances there is
little doubt that Man's consumption of energy at this stage was
only what was needed for bare survival, and that his existence
remained within the general context of a predatory animal until
remarkably recent times.

The first significant departure from an essentially 'wild' existence occurred only about 10,000 years ago. For the first time
in the history of our particular animal class, the mammals, one

particular species started to domesticate others and to cultivate a selection of plant species for its consumption as food. This was the agrarian revolution which took place towards the beginning of the so-called Neolithic period which commenced about 10,000 years ago. No doubt this development was made possible due to the advanced mental capacities of Man compared with other animals. The total energy that lay at Man's disposal – energy from cultivated plants, power from draught animals – increased immensely beyond what was available to Man the hunter.

How this transformation was accomplished remains obscure, but it appears to have happened at some time after the end of the last glaciation. Archaeological sites dated between 7000 and 6000 BC in present-day Iraq and Iran show clearly that people had at this time domesticated animals and grown crops such as wheat and barley. Similar evidence is found at slightly later dates around 5000 BC in South East Asia and China and around 4000 BC in Southern America. Between 4000 and 2000 BC the agrarian revolution spread into most of the present-day countries of western Europe, including the British Isles, derived it is thought from foci in the Middle East. With the dawn of the agrarian revolution also came the development of community life, with the emergence first of villages and then of towns. There was also the beginning of trade between communities with the concept that excess crops and animals could be exchanged or sold. The growth of prosperity and the general improvement in standards of living led to the first major human population explosion, leading in turn to an explosion in the global energy requirements of our species as a whole.

In the days immediately prior to the agrarian revolution the human population throughout the world would not have exceeded a few million. The reason was that only a limited number of animals were available to be hunted and also a limited amount of edible fruit and plants. If many more people than the hunt could feed were suddenly to be born, an excess would die of starvation, reducing the population back to an optimum level. The agrarian revolution changed all that. Crops and livestock came under Man's control. At the beginning of the Christian era the world population is believed to have been at about 150 million. By AD 1800 it had risen to 900 million. At this time the hunting life-style had been abandoned by most peoples of the world, notable exceptions still remaining in remote parts of Africa, South America and Australia.

With the development of agriculture to the level of a fine art, human beings suddenly found themselves freed for the most part of the purely animal struggle for survival that had existed in earlier times. Their new-found leisure was soon directed to higher intellectual pursuits. They developed skills in communication and writing, they painted pictures, and they probably spent an increasingly large fraction of their time in the abstract contemplation of the Universe. It is obvious that leisure in itself would not have sufficed or led to any notable achievement. An important prerequisite was that the human brain should already possess the capacity to undertake these pursuits and to make intellectual leaps forward. From these early human activities art, music, literature, mathematics and philosophy were undoubtedly born, as also our skills at constructing increasingly more sophisticated equipment and appliances to aid us in our daily chores.

In burning firewood for heat, growing crops, using wagons driven by other animals for transport, we were not using vastly more energy than was the lot of other animals of similar size. At most an average human might have used the energy consumed by, say, ten other animals of similar size. Even in present-day agricultural societies the per capita energy consumption for the most part does not exceed 10,000 kilocalories per day, on a scale where a sedentary worker might need to consume 2000 kilocalories. These are of course only average estimates, and there were exceptional instances where vastly more energy was consumed, as for instance in the building of the pyramids. For the latter operation a great deal more energy than 10,000 kilocalories per day per worker was harnessed, but again the basic source of this energy was animal muscle, which in turn was fuelled by plants that were consumed.

The first major departure from the use of biological energy came only at around AD 1800, with the advent of the industrial revolution, thereby unleashing forces that were hitherto unknown on our planet. The industrial explosion might be said to have begun in the early eighteenth century with the invention of the steam engine by Thomas Newcomen, although extensive commercial use of the steam engine did not get going till about 1820. Steam engines were used not merely for transport but also for extensive mining of coal and metal ores and in the textile industry. The energy source that fuelled the steam engine was of course derived from the burning of coal. Man at long last had learnt to harness not only the stored chemical energy in present-day life-forms but also the energy stored first in coal and later in oil. These so-called fossil fuels are in essence stores of solar energy that were accumulated by forests that covered the Earth during the carboniferous era about 300 million years ago. A life-force was thus harnessed in a way that connected present life with life that had existed vast spans of time ago.

If this was all the energy store that was accessible to Man, the dawn of the industrial revolution may have marked a fateful moment in the history of our species. By the year AD 1800 the production of coal worldwide amounted to some 15 million tons per annum. By 1950 it had risen to 1500 million tons per annum, with a steeply increasing demand for non-biological energy of this kind. Until very recently it was thought that the available store could not have supplied our escalating energy needs even for a few centuries, which would mean that an eventual return to a more primitive life-style would have become inevitable. There was a certain urgency defined in relation to our destiny: all that needed to be achieved had to be achieved before an inevitable energy crisis intervened. This crisis now seems to have been averted, for the time being at least, by the discovery of nuclear energy. But that discovery may not be without its problems. The escalation of the nuclear arms race and the stock-piling of nuclear weapons on both sides of the iron curtain presents an ever-increasing threat.

A possible way out of these difficulties was recently pointed out by the astronomer Thomas Gold. Gold has argued that a good deal more oil of extraterrestrial origin might lie buried in the Earth, coming from the days when our planet was being moulded by cometary and meteoritic impacts. We have known for a long time that carbonaceous meteorites contain significant proportions of hydrocarbons, and from last year's studies of Halley's comet we know that cometary material is almost as much complex hydrocarbons as water, and hydrocarbons are of course potential fuels. If comets are responsible for bringing the Earth's oceans, then oceans of oil must also lie buried in the Earth's crust.

According to the point of view we shall develop in later chapters, organic matter in comets is derived from biology, not from abiotic processes as Gold originally suggested. If this is so, our eventual future on this planet might be seen as contingent upon unlocking the stored energy of cometary life. The available energy store could well be measured in hundreds of thousands of years, still limited but amply long for other major developments to intervene. Human beings may by then have moved on to higher things.

TWO

The legacy of comets

Comets and the conditions within them are central to the arguments in this book. We have seen in the last chapter that cometary impacts played a major role in moulding our planet in its earliest days. More importantly, as it will turn out, comets are responsible for the origin and dispersal of life on a cosmic scale. For this reason we survey in this chapter the development of our understanding of comets in broad terms. We shall return to discuss matters relating to specific details as our argument unfolds in later chapters, but a general state of the art survey leading up to the most recent discoveries would now be in order.

Anecdotes and reports concerning the appearance of comets stretch back to the beginnings of recorded history, with archaeological evidence taking these records to a date that may even have predated the dawn of the agrarian revolution discussed in the last chapter. Reports of bright comets have certainly punctuated the chronicling of our history at various times. For instance, sightings of Halley's comet are noted in AD 1066, the year of the Battle of Hastings. It is clear that comets, whenever they have appeared, never failed to trigger the imagination of the beholder, invariably arousing a sense of inquiry that was directed towards probing their ultimate nature.

In common with many ancient commentators, the Greek philosopher Aristotle (384–322 BC) held that comets signalled disasters including plagues and pestilences. This view was so widespread and held with such tenacity across diverse cultures that one is tempted to speculate that causal relationships had actually been established over thousands of years. We shall return to this thesis and its consequences for the future of human society in a later chapter.

The Roman philosopher Lucius Annaeus Seneca (4 BC–AD 65) wrote as follows nearly 2000 years ago:

> Some day there will arise a man who will demonstrate in what regions of the heavens comets take their way, why they journey so far away from other planets, what their sizes and nature must be ....

Throughout the Middle Ages the view that comets journeyed to and from distant regions outside the Earth was not common. The most widely held belief was that cometary apparitions arose due to some meteorological effect that was confined entirely to within the Earth's atmosphere.

The first major milestone in the conquest of comets followed from the work of Edmund Halley (1656–1742). He found close similarities between the orbits of three comets which appeared successively in 1531, 1607 and 1682, with an interval apart of nearly 76 years. He concluded that these were appearances of one and the same comet moving in an oval-shaped (elliptical) orbit around the Sun and following the laws of motion enunciated by his friend Sir Isaac Newton. Halley correctly predicted a further return of this comet in 1759, an event that led to the general acceptance of the point of view that comets were objects that journeyed far away from the Earth in their own distinctive orbits around the Sun.

Attempts to understand the hidden source of splendour of comets, the so-called cometary nucleus, began in earnest with the work of F.W.Bessel (1784–1846) through studies again of Halley's comet during its apparition of 1835. He observed an intricate fine structure of jets, rays and fans emanating from the comet's head and developed the theory that solid particles (bacteria in our view) and molecules were being expelled.

In 1864 Giovanni Donati (1828–1873) used the science of spectroscopy to split the light from a comet's tail into its component colours. When he looked at Comet Tempel in this way he found that the spectrum showed three faint emission bands at particular wavelength ranges (colours) corresponding to radiations from molecules of diatomic carbon and cyanogen. Two years later William Huggins (1824–1910) observed a cometary spectrum in even greater detail. He confirmed the presence of the molecules seen by Donati, but in addition discovered the spectrum of reflected sunlight, thus proving the presence of reflecting dust particles (which could be bacteria and viruses) as well as gas in the material that was expelled from the nucleus. These were the foundations of cometary science which were laid already in the second half of the nineteenth century. Much remained to be discovered in the years between 1900 and 1986 as we shall now proceed to outline.

Today we know for certain that comets are linked inextricably with the destiny of the Earth. Cometary bodies, as well as their debris in the form of fine dust particles, have been colliding with the Earth from the very earliest days of its existence as a planet some 4600 million years ago. As we have already mentioned in chapter 1, it is now clear that volatile materials, including water, from comets contributed in significant measure to the primordial oceans and atmosphere. It is equally certain that a primitive Earth, which at the outset must have possessed a stark Moon-like surface, was transformed into a habitable form mainly due to contributions from comets.

The importance of cometary impacts and the acquisition of cometary material did not stop at a distant geological time, however. The entire solar system today is surrounded by a spherical halo of comets numbering some hundreds of billions and located at distances up to a light year or more away. Passing stars cause cometary objects from this halo to become deflected into orbits that take them to the inner regions of the solar system at the rate of about one or two every year. These deflected objects first show up as comets with long periods of revolution and a fraction of them later become rounded up by the gravitational pull of the massive planet Jupiter into orbits of shorter and shorter periods. Comet Kohoutek (1973) is an example of a long-period comet, whereas Comet Halley has a shorter period of approximately 76 years. There are comets of even shorter periods such as Comet Encke which has a period of 3.3 years.

The orbits of the known relatively bright short-period comets

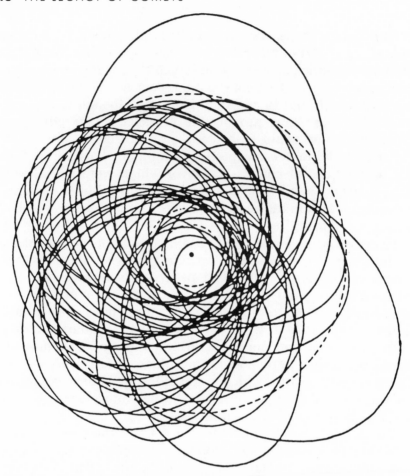

Fig. 2.1 Projections onto the ecliptic plane of the orbits of known short-period comets lying between the orbits of Mars and Jupiter. (Adapted from N.B.Richter, *The Nature of Comets*, Methuen, 1963)

are shown on the projection depicted in Fig. 2.1. The Earth is well and truly entwined within these orbits in a way that material shed from comets must surely be reaching us quite plentifully. Even the most conservative estimates point to a rate of picking up of comet debris of the order of some tons per day. A little of this debris has actually been collected in the atmosphere using high flying U2 aircraft and brought down for laboratory studies in recent years. Cometary material becomes trapped in the Earth's upper atmosphere, adding steadily to the store of terrestrial volatile materials. Fluctuations must occur in the rate of infall of cometary debris over long periods of geologic time. It has been suggested that significant increases in this rate could trigger the onset of episodes of global glaciations, the last of which ended nearly 11,000 years ago at a time which may have coincided with the earliest beginnings of the agrarian revolution, as we have already discussed.

A typical comet such as Comet Halley has a mass of some 100 billion tons, and a direct hit from such an object would have dramatic consequences for our planet. Fortunately such collisions are exceedingly rare, occurring with an average time separation of about 300 million years, approximately coinciding with the time interval between successive major bursts in the evolution of terrestrial life.

Smaller comets are more numerous and could collide more frequently. In 1908 a comet weighing some thousands of tons probably collided with the Earth in the Tunguska Valley of Siberia, exploding at a height of about ten kilometres in the atmosphere and causing devastation of a forest over hundreds of square kilometres. Already in this event a biological component was hinted at by several observers. Eye-witnesses stated that reindeer in the vicinity were stricken with mysterious diseases and many are said to have died. The Soviet biologist Dr Vasilieyev of Tomsk University, who led one of the expeditions to the site, wrote:

There have been the most violent genetic changes, not only in plants, but in small insect life. There are ants and other insects quite unlike those seen anywhere else or at an earlier time. Some of the trees and plants just stopped growing. Others have grown many times, many hundreds of per cent faster than they were doing before 1908.

So much is more or less agreed upon by most scientists. A less popular view advocated by us is that comets were also responsible for the importation of organic structures that contributed to the origin of life on our planet. We have further argued that the Earth continues to receive ready-formed living structures such as bacteria and viruses even at the present time.

From 1975 onwards we have accumulated evidence to support

our view that organic dust grains exist on a vast cosmic scale. These dust grains populate the clouds of gas that exist between stars and give rise to the visual effect of dark patches and striations seen against the diffuse light of the Milky Way. In 1981, through a combination of laboratory studies, mathematical computations and astronomical observation, we arrived at the conclusion that the bulk of cosmic dust was not merely organic but distinctly biological in character. We shall return to these ideas more fully in a later chapter.

There are arguments over a broad front that can be adduced to support the theory that life on Earth had its origins in comets and that evolution is controlled by the continuing input of cometary material. This input is not expected to be uniform with time because of the sporadic nature of the infall of comets from the outer solar system, and so the effect on evolution will also be sporadic. Such a sporadic character, often described as 'punctuated equilibrium', clearly shows up in the fossil record which traces the evolution of terrestrial life. A dramatic example is provided by the extinction of the dinosaurs 65 million years ago. It has recently been argued by many scientists that this event was caused by a comet that shrouded the Earth in a halo of fine particles. The particles acted like a smog that darkened the skies for decades and caused the withering of plankton and plants and the demise of all large creatures that fed on plants. We believe that such a purely physical cause is unlikely in view of the fact that the extinction of the dinosaurs also coincided with an extinction of a large fraction of the genera of all animals including microorganisms on the one hand, and on the other there was the emergence of several brand new orders of living forms. We identify this extinction/speciation event 65 million years ago with a genetic storm that resulted from an exceptionally large new crop of comets that were injected at this time into the inner regions of the solar system.

On 13 March 1986 cometary theories came under close scrutiny. The European spacecraft Giotto was set on a course that was to take it within 500 km of the core of Comet Halley, the comet that was depicted in the famous Bayeaux tapestry on its return in AD 1066, the year of the Battle of Hastings. The payload of Giotto contained a variety of scientific experiments designed to probe the nature of the material of Comet Halley, and also a video camera that was intended to send back colour pictures of its nucleus. Earlier in March the Russian spacecraft Vega 2 successfully accomplished a similar mission. From a distance of several thousand kilometres away the Russian pictures showed some tentative indications that the nucleus may be split into at least two fragments. Giotto might have been able to settle this matter in a spectacular way but for an unfortunate hitch in communications that occurred seconds before the moment of closest approach. In the event the best picture we have is a peanut-shaped nucleus surrounded by a shroud of obscuring dust (Fig. 2.2). Computer enhancements of such pictures, however, have led to claims that craters of sizes less than a hundred metres across are to be seen.

Pictures apart, what did the various missions to Comet Halley accomplish, and how do rival theories stand up in relation to the latest data? A theory put forward by R.A.Lyttleton in 1948 is that comets are aggregates of cosmic dust particles which have been rounded up by the gravitational pull of the Sun. Some degree of coagulation would occur to form larger-sized particles than cosmic dust grains, but in essence Lyttleton's model of a comet is one of a flying swarm of cosmic grains in orbit around the Sun. Data accumulated during the 1950s and 1960s made it difficult to maintain this model, at any rate in its original form. The main difficulty arose from the observation that gaseous material expelled from a comet as it approached the Sun was concentrated towards a sharp point that defined a presumptive core. Although a comet comprised of thousands of metre-sized pieces jostling with each other in close proximity within a small volume of space cannot be ruled out from general astronomical considerations, the data for Comet Halley seem to exclude such a model for this particular comet at least.

It was suggested in 1943 by Karl Wurm that comets consist of a collection of volatile ices, a model developed in later years by Fred Whipple, and known nowadays as the dirty snowball model. The idea is that molecules of various sorts, together with solid particles of dust, are trapped loosely within a matrix of

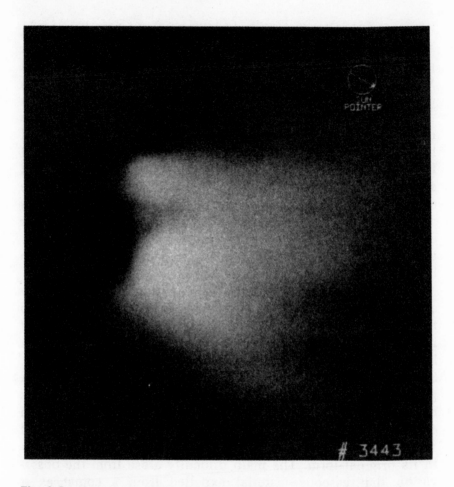

Fig. 2.2 Photograph of Comet Halley taken by cameras aboard the Giotto spacecraft. (Reproduced by courtesy of the Max Planck Institute for Aeronomie)

frozen ices – mainly water ice, ammonia ice and carbon dioxide. When the comet approaches the Sun the matrix is disrupted at the surface, releasing the trapped molecules, dust grains and particles of ice. In the case of Comet Halley the material peeled off at the 1910 approach to the Sun is estimated at about 200 million tons. A similar amount would have been peeled off during the present transit.

In the early 1970s it was discovered that the outflowing gas from a comet contains large amounts of the hydroxyl radical (OH) and the hydrogen atom (H). Because H and OH can add to H_2O, this observation was widely considered an unequivocal vindication of the dirty snowball model. Whilst it is obviously true that water is one possible source of these molecules, it is by no means the only source. Organic molecules can also act as a source of both H and OH, and it can be argued that this is more likely, for the reason that a number of carbon-bearing molecules and fragments of molecules, including C_2, C_3, CN, CH, had been observed in cometary environments for a long time. Moreover, in recent years searches for an infrared signature of water ice (at a wavelength of 3.07 microns) had consistently produced negative results. Thus the dirty snowball model might have been thought suspect even before the Giotto mission was planned.

The organic theory of comets described earlier began to emerge as a strong contender in the field immediately following the observations of Comet Kohoutek in 1973/74. The first clear-cut detections of organic molecules (hydrogen cyanide, HCN) and methyl cyanide (CH_3CN) came from this comet, and also a broad emission feature centred on the infrared wavelength of 10 micrometres. Although most astronomers were inclined to identify this latter feature with a spectral property of inorganic mineral particles, the way in which the infrared emission from the comet declined sharply just after its closest approach to Sun when the cometary dust became heated to 400 degrees Celsius indicated that it must rather arise from particles comprised of organic polymers.

The organic theory of comets predicts that cometary nuclei must develop a highly porous mesh-like surface layer of polymeric particles. Evaporation on a large scale can only occur from places on the surface where material becomes abrased or broken in some way. A consequence is that cometary nuclei must have exceedingly black non-reflective surfaces, a property we predicted immediately before the Giotto encounter. The Giotto pictures showed a few places where the nucleus could be viewed through 'holes' in the surrounding dust shroud, and the surface was seen to be amazingly black. Giotto investigators described it as being 'blacker than the blackest coal', and this was indeed a veritable triumph for the organic model. More triumphs were to follow. Giotto's particle impact analyser was equipped to determine the chemical make-up of the cometary dust particles, by measuring the distribution of the masses of atoms within them. J.Kissel, the principal investigator on this experiment, reported on 17 March that analysis of 1 per cent of the total data record indicated that a considerable fraction of the dust was made up of the elements carbon, oxygen, nitrogen and hydrogen, with very much smaller amounts of other chemical elements. There are two logical possibilities that follow from these data. These atoms can exist in the form of volatile ices such as water, carbon dioxide and ammonia, together with simple hydrocarbons such as methane, or they could occur in the form of highly stable organic polymers. Since in other experiments the temperature of these particles was measured at about 50 degrees Celsius, well above the evaporation temperatures of simple hydrocarbons and inorganic ices, it follows that the fraction of cometary particles in question could only be made of organic polymers. Furthermore, the sizes of the particles and their densities determined in Giotto experiments make them fully consistent with a model involving dry microorganisms.

Both Vega 2 and Giotto experiments indicated that hydrogen atoms, hydroxyl radicals and water molecules were also released quite liberally from the comet. However, the amounts of these substances were not inconsistent with the water associated with living cells (more than 60 per cent by volume of such cells is water). Significantly there was no evidence whatever for a halo of icy particles around the nucleus which had been predicted by the dirty snowball model.

Scarcely two weeks after the Giotto rendezvous, an important observation of the comet was made in the infrared region of the spectrum by D.T.Wickramasinghe and D.A.Allen. These astronomers used the 154 inch Anglo-Australian telescope at Siding Springs Mountain in New South Wales. They discovered remarkably strong signals due to emission from heated organic dust over the wavelength range from 2–4 micrometres. Basic structures of organic molecules involving linkages between carbon and hydrogen atoms (CH bonds) absorb and emit radiation at wavelengths near 3.4 micrometres, and for any assembly of

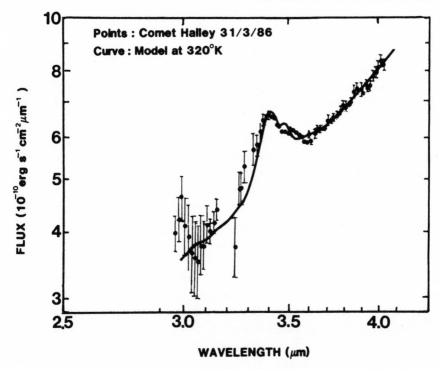

Fig. 2.3 Infrared spectrum of Comet Halley taken at the Anglo-Australian
Observatory by David Allen and Dayal Wickramasinghe (points) compared
with the normalised spectrum of dry bacteria (curve)

complex organic molecules such as a bacterium this absorption
band is in general very broad and takes on a highly distinctive
profile. The most remarkable fact that emerged was that the
profile of this emission from Halley's comet matched very closely
the behaviour of a dried bacterium as measured in the laboratory.
This correspondence is shown in Fig. 2.3.

Although a full appraisal of the recent observations of Comet
Halley cannot yet be offered, it is already clear that a number
of quite remarkable facts have emerged. The comet is clearly
not an inorganic dirty snowball as most astronomers had held
against all the odds. It is in a considerable measure organic.
Within a mammoth chunk of organic matter some 8 km long
and 4 km wide a large proportion of water molecules and traces
of other materials are also trapped. Whether one likes it or not,
the organic material of the comet occurs predominantly in the

form of particles whose absorption properties, sizes and densities are identical to the predictions of a bacterial model.

We have referred already to the extreme blackness of the surface of Comet Halley. The surface is not black because of absorption of sunlight in the immediate surface material, for if there were almost complete absorption close to the surface, the subsurface material would not be much heated and would not be subjected to explosive evaporation processes such as were observed. Surfaces do not need to be absorbent in order to be exceedingly black. Materials reflect sunlight according to the extent to which their refractive indices are disordered on the scale of the wavelength of light. Ice can be either very bright or very black according to whether or not it contains myriads of microscopic-sized air bubbles. Take a sample of bright bubbly ice, melt it to get the air bubbles out, and then refreeze and you have black ice. The facts suggest that the surface material of Comet Halley is not disordered with respect to refractive index, and if the surface material has porosity, as may well be the case, the pores must be systematically arranged as within a silica gel, not randomly placed as with the air bubbles in bright ice.

It is common for materials that are translucent at one waveband to be opaque (i.e. absorbent) at other wavebands. If cometary surface material is highly translucent at optical wavelengths but highly absorbent at infrared wavelengths around 10 microns, then the surface layer, which might be several metres thick, would act to produce a strong greenhouse effect. Energy from sunlight would go in, to be eventually absorbed at depth, say at depths of 10 to 20 metres, as is the case for black ice. Once the sunlight was converted to heat there would be an impediment to the heat escaping back out into space. The energy from the Sun would thus become stored, raising the temperature of subsurface material.

Rising temperature could produce melting, with ensuing chemical reactions then occurring in the subsurface material, but except on the unlikely supposition that the material is itself inherently explosive, non-biological chemistry would be a pretty mild affair. Biochemical reactions depending on enzymes are, on the other hand, millions of times faster than ordinary reactions. Comet Halley has been found to rotate in a time exceeding 50 hours, and a bacterial culture that is permitted to grow for an interval as long as that can do a great deal, especially if there is a gaseous output from the enzymic reactions, as is commonly the case. Gas accumulating to sufficient pressure would lead to explosive outbursts from below, penetrating through the surface material like leaks from a water main, causing evaporating liquids and suspended bacteria to be spurted out into space, which is exactly what can be seen from observation to happen.

The curious comet Schwassmann-Wachmann I moves, not in the usual highly elliptic form of cometary orbit, but in a nearly circular orbit with radius a little larger than Jupiter's orbit. About once in 15 years there is a violent outburst from this comet comparable in scale to outbursts from Comet Halley. Because of the low intensity of sunlight at and beyond Jupiter's distance from the Sun, there is no possibility that the outbursts of Schwassmann-Wachmann I can be due to ordinary thermal evaporation such as is assumed by protagonists of the dirty snowball model. Provided, however, that the effect of sunlight is sufficient for a bacterial reaction to 'go' then, even at a gentle rate of production, gas would still accumulate to the point of explosive outburst. The case of Schwassmann-Wachmann I has always given a clear demonstration that the dirty snowball model must be wrong, but it fits the bacterial model without difficulty.

For a long time cometary science appeared to have been in the doldrums, an esoteric subject with little relevance to the rest of astronomy, let alone to the rest of science. This situation has changed over the last two decades. The latest discoveries interpreted in the most conservative possible way indicate that at the very least, comets provided the raw material, the complex organic building blocks, from which life emerged on the Earth. More realistically, in our view, the data point in a decisive way to life in the form of bacteria, viruses and even larger more complex cells, being actually contained in comets.

THREE

The conquest of the
solar system

There was a time earlier in this century when the solar system, consisting of our central star, the Sun, its nine planets and their satellites, very many minor planets and the 100 billion strong cometary cloud which surrounds it all, was thought to be unique. Such a point of view was a consequence of the so-called 'tidal theory', which maintained that a passing star came so close to the Sun as to pull out a cigar-shaped filament of gas from its surface out of which the planetary objects subsequently condensed. The course of events involved in such a model was estimated to be so exceedingly improbable as to imply that we could be part of a planetary system which must essentially be unique in the galaxy.

Nowadays many arguments can be used to dispose of this model and it can be taken for granted that the formation of a planetary system, far from being a rare event, must be a relatively common one. The formation of planetary systems seems part and parcel of the process by which new stars are formed from the collapse of cosmic dust clouds. Stars are seen to form in prolific numbers wherever there is cosmic gas and dust intermingled with fully fledged stars. New-born stars such as are known to exist in the Orion nebula are immersed in cocoons of dust which radiate energy mainly at long infrared wavelengths. Many of these objects are scarcely a million years old and some new stars have not even had time to switch on their nuclear sources of generation of energy. Even earlier stages in the formation of stars are becoming accessible to observation, thanks to advances in infrared astronomy. Many sources of infrared radiation recently discovered by IRAS, the Infrared Astronomical Satellite, are compact clouds of interstellar material which are on the verge of becoming star/planet systems. One particular star, beta Pictoris, has been found to have a flattened disc of material around it which is well on the way towards becoming a planetary system.

The solar system began its life as a fragment of a cloud of gas and dust that broke off and collapsed from a larger complex of interstellar material such as occurs in great profusion throughout the Milky Way. In a later chapter we shall show that this cloud must already have contained the evolutionary potential for all life in the form of bacteria, viruses and larger cells such as yeast, the inheritance of life with which the solar system was initially endowed. A small initial rotation of the break-away cloud fragment was amplified as the collapse proceeded, eventually leading to a rapidly spinning central star which was itself capable of flinging off a disc of hot gas due to its fast spin. It was from this disc of hot gas that we believe the inner planets condensed. The locations of the inner planets in their present-day orbits are consistent with such a model, where materials of different volatilities condensed from the cooling gas at different distances from the primitive Sun.

Not all the gas in the original cloud fragment falls in to form the central star, however. The outer part of the cloud remains cool enough to accumulate eventually into many billions of frozen cometary bodies, a typical body measuring a few kilometres across. This, in our view, is the beginning of the Oort Cloud of comets. A fraction of the initial small cometary bodies must have coagulated further to give rise to the outer planets, Uranus and Neptune, but the rest would remain in an essentially pristine state. Objects within the Oort Cloud are expected to have accumulated radioactive isotopes such as Al_{26}, which has a half-life of about a million years, and this material will initially provide an internal heat source. Whatever radioactive heat was released after the comets condensed would have led to a melting in their interiors, and to the maintenance of an internal core of organic materials and water under lukewarm conditions for some millions of years. It is this heat source that is ultimately responsible for the amplification and development of the cosmic life-force at the time of the inception of the planetary system. The initial legacy of viable cells and viruses that was incorporated into the original comet cloud may well have been considerable, but it is not required to be so, because lukewarm interiors of comets which were maintained for a considerable span of years permitted a vast amplification of the original component of life that the planetary system would have contained.

The essence of living systems is that they reproduce. Be it mammals, birds, insects or bacteria, the most apparent manifestation of the life-force is a tendency to increase the numbers within individual species, thereby transforming the inorganic environment around these life-forms to organic form. It is our view that wherever organic material is to be found in quantity biological processes have been at work. The organic matter produced through biology will persist even after the plants and animals that processed it have perished, until eventually the organic material becomes oxidised or scavenged by other life-forms.

All the organic material on the Earth today is either directly or indirectly due to biology, and so it is likely to be for organic matter that is found in substantial quantities elsewhere in the Universe. In this context even the existence of methane may be indicative of biological activity in the solar system. Astronomers have been content to accept the methane observed in the atmospheres of the four large outer planets of the solar system as the outcome of a 'thermodynamic trend' of carbon compounds to go to methane at low temperature and in the presence of an ample supply of free hydrogen. But if one takes a container with a mixture of free hydrogen and carbon monoxide or carbon dioxide the separate gases would persist unchanged for an essentially infinite length of time. The trend is so slow as to be essentially zero. It is in such situations that catalysts are used in the laboratory and in industry. Of all catalysts in the Universe bacteria are the most efficient. Indeed, the methane-producing bacteria exist precisely by speeding up the conversion of carbon dioxide and hydrogen to methane and water. If the conversion happened at all readily under inorganic conditions there would be no niche for this particular kingdom of bacteria.

Let us look elsewhere in the solar system, outside the Earth, for signs of cosmic life. If comets carry microorganisms as we have proposed, and if it was from comets that Earthly life was derived, one would expect to see signs of life on other planets and satellites within the solar system. Fortunately a fraction of these objects have been explored in recent years, albeit by remote sensing of one kind or another.

The nearest celestial neighbour that presents itself is of course the Moon. But because of its small size compared with that of the Earth and the consequent lack of an atmosphere, extremes of temperature at the lunar surface and the generally inhospitable environment that it presents, the case for any life-forms surviving there must have seemed slim even a century ago. Yet science fiction writers were never lacking in their imaginative feats of creating scenarios for the survival even of intelligent life on the Moon. Thus Sir James Jeans wrote in 1942 (*Science*, vol. 95, p. 589):

> In 1829 a New York newspaper scored a great journalistic hit by giving a vivid, but wholly fictitious, account of the activities of the inhabitants of the moon as seen through the telescope recently erected by His Majesty's Government at the Cape.

That a joke of this kind was even contemplated is an indication of Man's natural desire for detecting cosmic life outside our own planet. Yet Jeans' pessimism for being able to verify the truth or otherwise of such statements was unwarranted, as it eventually turned out. Jeans went on to say:

> It will be a long time before we could see what the New York paper claimed to see on the moon – bat-like men flying through the air and inhabiting houses in trees – even if it were there to see. To see an object of human size on the moon in detail we would need a telescope of from 10,000 to 100,000 inches aperture, and even then we should have to wait years, or more probably centuries, before the air was still and clear enough for us to see details of human size ...

What Jeans could not have anticipated, of course, was the priority accorded to rocket engineering towards the end of the Second World War and the eventual development of rocketry capable of launching spacecraft for interplanetary travel.

But before we come to describe the benefits reaped through space travel, let us first recount a somewhat romantic tale of Man's encounter with the notorious red planet, Mars. In many ways Mars is not too dissimilar to the Earth, and might have been thought to come nearest to conditions prevailing on Earth. About half the size of the Earth and about 50 per cent more distant on the average from the Sun, Mars is clearly made of Earth-like stuff. At its closest approach to the Earth it comes to within a cosmically small distance of some 35 million miles. Mars has a thinner atmosphere than the Earth, but one that even now could give adequate protection for some primitive life-forms to survive. Mars has seasons just like the Earth, but they are longer, and its year is longer too, measuring some 687 Earth days. The Martian day, on the other hand, is not very different from an Earth day, measuring 24 hr, 37 min, 22.7 sec. Mars has two irregular-shaped satellites, Phobos (27 km along its major axis) and Deimos (15 km in its major axis), circling around it almost in the same plane. There have been suggestions, half serious perhaps, that these satellites might be artificial constructs created from material resembling carbonaceous chondrites (carbon-rich meteorites).

Man's search for extraterrestrial life might be said to have begun in earnest in 1877 with a report by the Italian astronomer Giovanni Schiaparelli that he had seen unmistakable evidence of long straight-line features on the Martian surface. The legend of the Martian canals was thus born. The implication was of course that these were artificial constructs of an intelligent life-form that was thought to inhabit the planet. The search for canals on Mars was taken up more systematically by the American astronomer Percival Lowell, who was wealthy enough to build his own observatory and telescope in Flagstaff, Arizona, which for a long time was to be dedictated to this very pursuit. Lowell was soon to announce that his observations showed an indisputable network of canals spreading outwards from the polar caps, the seasonal melting of which was thought to supply water across the 'continents' of this arid planet. Percival Lowell wrote thus:

Fine lines and little gossamer filaments only, cobwebbing the face of the Martian disk, but threads to draw one's mind after them across the millions of miles of intervening void ....

This great number of lines forms an articulate whole. Each strand joins to the next (to the many next, in fact) in the most direct and simple manner – that of meeting at their ends. But as each has its own peculiar length and its special direction, the result is a sort of irregular regularity. It resembles lace tracery of an elaborate and elegant pattern, woven as a whole over the disk, veiling the planet's face. By this means the surface of the planet is divided into a great number of polygons, the areolas of Mars.

Fig. 3.1 shows a depiction of Lowell's lace network of canals. Around the turn of the century Lowell had published a series of books in which he proposed the existence of advanced civilisations on Mars to account for these so-called canals. His writings served to inspire science fiction novelists like H.G. Wells to write stories that surprisingly enough very often terrified the reader, and in more recent times such novels gave way to movies such as ET.

A final resolution of the Martian life question became possible only after the dawn of the 'space age'. The earliest close-up pictures of Mars taken by cameras aboard the Mariner spacecraft in the 1960s cast very serious doubts on the existence of any structures even vaguely resembling the canals of Lowell. The implication was that the canals of Schiaparelli and Lowell were no more than optical illusions: the human brain tending to line up indistinct features such as were present in considerable abundance in images of the red planet. It was, however, not until 13 November 1971, when Mariner 9 went into orbit around Mars and sent back some 6900 images, that the spectre of Martian intelligence was finally laid to rest. The absence of canals was then confirmed in a decisive way.

This still left open the possibility of lowly life-forms existing on Mars. The best prospect for biological activity on Mars is probably inside glaciers, where it may be possible for temperatures to rise sufficiently for liquid water to exist. Bacteria in such a situation would need to live as 'chemotrophs', which is to say bacteria which live on the energy produced by some chemical reaction. Many bacteria in such conditions produce

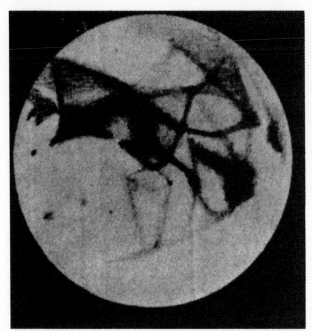

Fig. 3.1 A drawing of Lowell's canals on Mars compared with an actual photograph taken in 1926. We now know that these markings do not correspond to any physical features on Mars

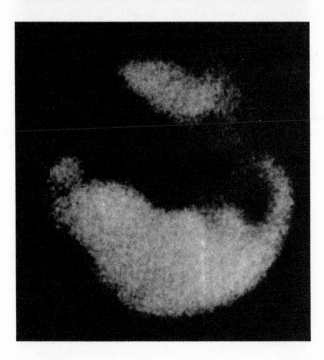

inorganic gas as a product of their metabolic activity. If such bacteria were present, subsurface pockets of gas would be expected to build up. When gas pressures rose to a point that could not be contained by the overlying ice, the surface would break open, leading to the explosive release of vast quantities of bacterial particles.

It seems significant in this context that a massive dust storm occurred at the Martian surface just at the time when Mariner 9 arrived there in 1971. Spectra of the light reflected from the dust showed patterns similar to that which would be obtained from bacteria. Such dust storms on Mars are traditionally attributed to the effect of high winds generated by the Martian climate, but if this were so it is hard to understand why the storms should not occur periodically with the seasons. A possible, and perhaps even the most likely, explanation of the 1971 dust storm is the explosive release of subsurface bacteria due to the build-up of pockets of high gas pressure.

Although liquid water is not present now at the Martian surface, sinuous channels have shown up in many close-up pictures, indicating that the channels have been cut by the flow of liquid water. The occurrence of liquid water on the Martian surface at some earlier epoch seems a distinct possibility that is consistent with all the available evidence. In the present day Mars has only traces of water in the atmosphere, with larger quantities of carbon dioxide (CO_2), carbon monoxide (CO) and nitrogen (N_2). With such an atmospheric composition, advanced life-forms resembling terrestrial plants and animals could be ruled out, but lowly life in the form of bacteria, fungi and lichens is a possibility that clearly needs to be explored.

To this end the US space agency NASA planned two unmanned missions to Mars: Viking 1 arrived on 20 July 1976 and Viking 2 on 3 September 1976. Each mission involved an 'orbiter' that was set in orbit around the planet and a 'lander' that actually landed at a chosen spot, the prime objective being to search for life on Mars. The Viking landers arrived safely at predetermined locations on the Martian surface, equipped with apparatus to make in situ tests for microbial life.

In one experiment (designated LR) a nutrient broth of the sort that is normally used to culture a wide range of terrestrial bacteria was contained in a sterilised flask, and Martian soil was added to it. It was found that the nutrient was taken up by the soil and gases were expelled from the flask, as would be expected if bacteria were present. In another experiment the soil sample was heated to 75 degrees Celsius for 3 hours before it was added to the nutrient. This led to a diminution of the gas release by 90 per cent, but significantly the reaction was not completely stopped. Since some bacterial and fungal spores could survive temperatures of 75 degrees Celsius, the result of this second experiment was also consistent with a biological explanation, especially as the activity recovered gradually to its former higher value as time went on. The bacterial explanation gained further support from a third result, obtained by heating the soil sufficiently to kill microbial life entirely, when all activity was found to be stopped. However, yet another experiment in the Viking package proved initially more difficult to reconcile with biological activity on Mars. This experiment, designated GCMS, sought to analyse the organic content of Martian soil. Here the results were disappointingly negative for organic matter, indicating that if such matter existed it was present only in small quantities.

The fact that the LR experiment was decisively positive and the GCMS experiment was negative posed a difficulty for those involved in the NASA missions. The outcome was indefinite, and this is the way it should have been presented to the public. Yet NASA elected in 1976/77 to be cautiously overpessimistic in declaring that the Viking experiments did not support the presence of microbial life on Mars, even though the most crucial biological experiment (LR) had yielded a positive result. It was their view that some non-biological explanation had to be sought and would eventually be found. NASA investigator Horowitz wrote as follows, about a situation that must have given much cause for disquiet:

... it is not easy to point to a non-biological explanation for the positive results. Investigations into the problem are now under way in terrestrial laboratories with synthetic Martian soil formulated on the basis of data from the inorganic analyses carried out by the Viking landers. The solution to the puzzle will probably also explain why the organic-analysis experiment detected no organic material in the Martian surface. Until the mystery of the results from the pyrolytic-release experiment is solved, a biological explanation will continue to be a remote possibility ...

Ten years on and after extensive experimentation with simulated Martian soil under appropriate conditions, the Viking investigators G.V.Levin and P.A.Straat concluded that the balance of evidence favoured the presence of life on Mars. It has turned out that no non-biological model is feasible for explaining the positive results of the LR experiment, and moreover that the lack of free organic matter in quantity can be very easily explained on the basis of a slow turn-over rate of microorganisms to be expected under the inhospitable conditions that prevail on the planet. Indeed, the Viking experiments when they were subsequently tested on soil from the dry valleys of the Antarctic produced results virtually identical to those obtained on Mars. And since we know that populations of microorganisms exist in the Antarctic soil, they certainly cannot be excluded from Mars. Levin and Straat's conclusion in 1986 is that the Viking missions are more likely to have discovered life than not some ten years ago.

Another startling revelation reported by Levin and Straat in 1986 concerned a series of colour pictures of a Martian rock field obtained by cameras aboard one of the Viking landers. A succession of pictures taken at intervals through the Martian seasons showed a distribution of greenish patches that changed with time. The changes were strikingly similar to effects seen for distributions of lichens in terrestrial rocks. Lichens are known to survive under exceedingly dry conditions and are capable of thriving by taking up small quantities of water vapour that are present in the air. An interesting aspect of this discovery is that lichens are not the most primitive of Earthly life-forms. They are in fact a symbiotic association of fungi and algae, considerably more advanced than bacteria.

Other explicit searches for signs of life on extraterrestrial objects were first carried out for a class of meteorite known as the carbonaceous chondrites. As their name implies, the carbonaceous meteorites contain carbon in concentrations upwards of 2 per cent by mass. In a fraction of such meteorites the carbon is known to be present in chemical compounds of high molecular weight involving also the elements hydrogen, oxygen and nitrogen.

Although there is still some debate on the matter, it is generally held that carbonaceous meteorites are of cometary origin. Even though microorganisms are abundant within comets according to our point of view, meteorites do not at first sight seem promising as objects in which to search for life. This is because meteorites are essentially the end products of comets after all their volatile components have boiled off as a result of successive plunges into the inner regions of the solar system. One might think that such a process would have removed all traces of life from the original cometary body. It is also possible, however, that some sedimentary accumulations of bacteria could occur within the fast-shrinking comet and bacterial fossils might therefore conceivably be found in such accumulations, appearing eventually in meteorites landing on the Earth.

Microfossils of bacteria are remarkably indestructible and already in the 1930s there was a suggestion that fossilised spores were present in the Orgeuil meteorite that fell in 1864. In the early 1960s this speculation was transformed into a serious proposition by G.Claus and B.Nagy, who claimed that certain organised elements in the Orgeuil and Ivuna meteorites (the latter fell in Tanzania in 1938) were of biological origin.

The claim of Claus and Nagy provoked instant uproar. Rumours of contaminations within the meteorites circulated to an extent that proved highly embarrassing to these investigators. Yet the situation was by no means clear-cut. Although some contamination was involved, not all the organised elements found within these meteorites could necessarily be dismissed as contaminants.

This problem was taken up again early in 1980 by H.D.Pflug. In a long series of experiments carried out with great care Pflug found a similar profusion of 'organised elements' in thin sections prepared from a sample of the Murchison meteorite, a carbonaceous chondrite that fell in 1969 about a hundred miles to the north of Melbourne in Australia. The method adopted by Pflug was to leach out the great bulk of the minerals present in thin sections of rock, thereby permitting the insoluble carbonaceous residue to settle undisturbed on a piece of film. The sediment was then examined in a non-destructive way using an optical microscope, and also deploying the techniques of electron microscopy. The structures that are revealed are remarkably similar to certain types of bacteria and viruses as seen in Fig. 3.2. They are also similar to fossilised microorganisms found in Precambrian rocks. Attempts to explain these structures by non-biological processes do not seem to us successful, for the reason that in every case we have seen of small particles being produced by non-biological means the particles had the evident appearance of artifacts.

Denials of fossils in carbonaceous meteorites remind us rather strikingly of the earlier denials of the existence of meteorites themselves. Reports of meteorites falling from the skies stretch back to biblical times and are often well documented. Yet throughout the seventeenth and eighteenth centuries learned societies such as the French Académie des Sciences vehemently denied the evidence, maintaining that stones of extraterrestrial origin could not exist. It was only when a fall of over 2000 'stones from heaven' occurred at L'Aigle on 26 April 1803 that the Académie changed its opinion, and then only because it was impossible to deny such a prodigious fall seen by a multitude of people.

The first celestial object to which attention was directed in both the American and Russian space programmes was the Moon. With many hundreds of kilogrammes of lunar samples returned to Earth from 1969 onwards, the scope for looking for evidence of microscopic life landing there might have been thought to be good. Although microorganisms of cometary origin would inevitably be incident on the lunar surface, prospects for their survival as detectable units are grim. Microorganisms could not land intact upon the Moon because of the lack of a lunar atmosphere to provide a soft landing. Microorganisms hitting the surface of the Moon at speeds of about 30 km per second would be exploded into atoms and relatively simple molecules. The only feasible evidence that can be hoped for in the case of the Moon must therefore depend on gross elemental abundances, or on the presence in the lunar soil of very small concentrations of simple organic molecules that have managed

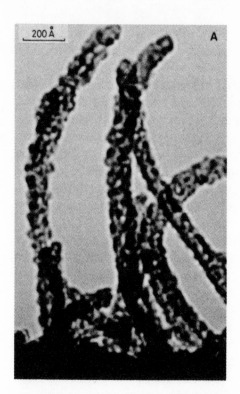

to survive the high-speed impacts of their parent organisms. It may be significant that small quantities of amino acids in samples of lunar soil have persistently been reported, while the case of carbon seems also significant. Elemental abundances in the lunar soil are mainly dominated by oxygen, magnesium, aluminium, silicon and calcium, just the elements that are dominant in terrestrial rocks, showing that lunar soil is made of rock-like mineral material. It is therefore curious that carbon, which is not a constituent of typical rocks, is present in lunar soil in a concentration of about one part in a hundred thousand. A rain of microorganisms from cometary sources is one source of carbon as well as of amino acids found in lunar soils. Another possibility for the lunar carbon is that this element was implanted from a diffuse flow of atoms that constantly blows outwards from the Sun. The fact that the carbon abundance is unaltered as one goes a metre or more into the surface causes

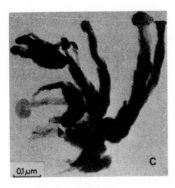

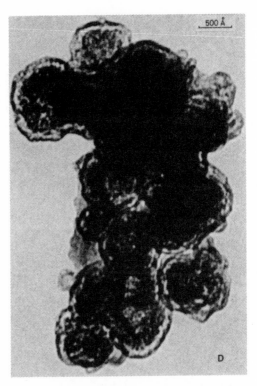

Fig. 3.2 Organic particles strikingly similar to known bacteria and viruses extracted from the Murchison meteorite by Hans D. Pflug. (A) structures generally similar to TMV particles ; (B) and (C) structures similar to the bacterial genus known as pedomicrobium ; (D) a cluster resembling a collection of influenza viruses

difficulty for the solar wind theory, however, which would account for an appreciable carbon abundance only for a thin skin of lunar material.

Beyond the Moon and Mars many other planets and satellites of the solar system have also been explored to varying degrees of thoroughness. In general there seems to be no dearth of evidence that could be interpreted as signs of the cosmic life-force. The physical conditions prevailing in 'other worlds' of the solar system span a wide range with respect to temperature, pressure, composition and the exposure to sunlight. The characteristics of bacteria that enable them to survive are also known to be so widely ranging that some bacterial types may exist to fit the environmental conditions in many of these 'worlds'. We take as an indispensable requirement for the replication of bacteria that water can condense on them and pass to their interiors to be held there in liquid form. This condition rules out Mercury and the Moon. The lifeless states of Mercury and the Moon may be seen immediately in their drab lack of colour. It is tempting to connect life and colour as being associated features connected with the surface properties of planetary surfaces, although this correlation need not of course be rigorously valid. A stricter indication of life comes when solid particles of sizes close to 1 micrometre are seen to dominate in any given environment. The size 1 micrometre is significant because it is the average diameter of spore-forming bacteria, whereas such a size could arise only accidentally in a freely-condensing non-biological system.

Spacecraft have visited the planet Venus from 1962 onwards and at the present time a wide range of information relating to this planet is available. Since Venus is exceedingly hot at ground level (about 450°C) it would be impossible for life as we know it to exist there. Venus, however, has an extensive cloud cover and it is within such clouds that life might have taken root. Water is present in regions of the atmosphere. Although water is not abundant, the higher atmosphere of the planet is cool enough for the vapour pressure of water to approach saturation. The Venusian atmosphere has a dual circulation pattern. The clouds of Venus are in convective motion in the upper atmosphere, which ranges in height from about 70 km to 45 km. The temperature at the top of this layer is about −25°C and at the bottom it is about 75°C. Below is a highly stable slice of atmosphere from 45 km down to 35 km, which separates the upper clouds from the much hotter regions below. Survival of bacteria over the range of conditions in the upper Venusian clouds would seem to be possible. The repeated variations of temperature caused by a circulating cloud system, however, tend to favour bacteria capable of forming spherical spores which are still more hardy than the bacteria giving rise to them.

The upper clouds of Venus produce a rainbow, indicating that the cloud particles are mainly spherical and that they have sizes in the region of 1 micrometre. Measurements of the refractive indices of these particles are also fully consistent with the properties of bacterial spores. The populations of cloud bacteria are expected to be periodically topped up by fresh cometary injections. Such injections could also supply the Venusian atmosphere with a steady input of inorganic nutrients to replace what must inevitably fall into the atmosphere below.

The NASA missions Voyager 1 and Voyager 2 undertook what was perhaps the most ambitious and thorough investigation of the solar system. Having first explored Jupiter and its satellite system, it proceeded outwards to return dramatic pictures of the Saturnian system, then of the Uranian system and now it is hurtling off towards Neptune. Major surprises unfolded at each stage.

Even before the discoveries of Voyager, indications for life on Jupiter, Saturn, Uranus and Neptune were in our view available. The large quantities of methane present in all these planets are readily explained as the action of methanogenic bacteria (methane-producing bacteria) which reduce carbon monoxide and carbon dioxide. For reasons stated earlier, it would be well nigh impossible to convert carbon dioxide and hydrogen existing initially in these regions into methane and water by any other process than biology.

We have argued that such features as the Great Red Spot on Jupiter are indicative of biology actually controlling the meteorology of this planet. A kilometre-sized object hitting the planet at high speed would be disintegrated into hot gas that would form a diffuse patch similar to the Great Red Spot. Such a region of atmosphere would be rich in the inorganic nutrients needed for the replication of biology. A large bacterial population could then be built up in this area.

The possibility exists for a feedback interaction to be set up between the properties of a localised bacterial population and the general meteorology of the Jovian atmosphere. If bacterial populations on Jupiter have become adapted to the meteorology there, it is possible that evolution has produced a situation in which populations are able to prevent supplies of nutrient materials from being swept away from them by atmospheric motions. In a measure they may have become able to control the meteorology, and thus to hold together spot concentrations of nutrient material. This may well be an explanation for the persistence of spots, and in particular the remarkable persistence of the Great Red Spot.

The volcanic activity found extensively in the Galilean satellites of Jupiter appears to point to bacterial populations at work underneath a surface crust. Bacteria that release gaseous prod-

Fig. 3.3 Voyager picture of the Jovian satellite Europa exhibiting a curious criss-crossing network of cracks in the surface ice. A brownish-coloured pigment appears to have diffused outwards from the cracks

ucts through their metabolic processes can produce accumulations of pockets of high gas pressure which periodically explode through surface cracks. The process is exactly similar to what we discussed in relation to the Martian dust storms. The surface of Europa is criss-crossed with an intriguing network of cracks in a surface ice layer (Fig. 3.3). The cracks have a mysterious brownish colour suggestive of a biological pigment. Antarctic sea-diatoms (a group of brown algae) seem prime candidates for organisms that can survive the subsurface conditions on Europa.

Micrometre-sized particles in the solar system are seen most spectacularly in the rings of Saturn. Astronomers were astonished when the Voyager missions first revealed the immense multiplicity of rings – a few had been expected, but nothing like the many hundreds of small ringlets that were actually found. Embedded in the ringlets were a considerable number of hitherto unknown small satellites which came to be referred to as shepherd satellites because it was thought that the intricate ringlet structure was due to a subtle gravitational control exercised by these hitherto unknown bodies. The idea is that an initially uniform disk of small particles revolving around Saturn would break-up into many ringlets as the small bodies were added by planetary capture. Whilst this is undoubtedly a possibility we think there is considerably more to the matter. The shepherd satellites are extraordinarily like the nuclei of comets, from which one may suspect that the small particles have been expelled from the satellites in the manner of the recently observed outbursts from Halley's comet. It is hard, however, to see how a planet like Saturn could capture comets under present-day conditions. Inevitably one seems led to the conclusion that the shepherd satellites were acquired at the time the planet itself was formed from cometary objects. Similar considerations would apply to the rings around Jupiter and Uranus which have been discovered only in recent years.

FOUR

The conquest of
our galaxy

The 'emptiness' of space is a grand illusion. The vast tracts of space that separate stars and groups of stars in our galaxy are not empty at all. They are occupied by material, which by all terrestrial standards is of course in a tenuous, dilute form. A cubical volume of space with a side measuring a thousand kilometres will on the average contain only one gramme. The interstellar material exists in many different forms: single atoms, combinations of atoms in the form of radicals and molecules, and tiny dust grains whose sizes are much less than the dimensions of a pin head.

From the mid-1960s onwards it became apparent that the molecules in space included a large component that was organic. That is to say, the molecules were made up of combinations including carbon, hydrogen, oxygen and nitrogen that are commonly associated with living things. Examples were hydrogen cyanide, formaldehyde, methanamine, all molecules that scientists had come to believe might be a necessary prerequisite for an origin of life itself. Equally well these molecules could of course be degradation products of living matter, and from the astronomical data alone it was not possible to tell which of these possibilities is correct. The question came to be resolved only when we began to probe with care and rigour the properties of the interstellar dust particles which are known to populate the interstellar clouds.

The nature of the cosmic dust has been a subject of controversy and discussion for a long time. Until the 1930s even the existence of interstellar dust was a matter for vigorous debate. Perhaps the most striking evidence for the existence of interstellar dust is still to be found in photographs of star fields in the Milky Way. They reveal very clearly the presence of dark patches and striations against more or less uniform distributions of stars. Fig. 4.1 is a typical example of such a photograph, showing the Milky Way in the region of Sagittarius with the central part of the galaxy obscured by dust clouds.

From 1940 to about 1960 astronomers held the view that this dust was made mainly of icy material, rather similar perhaps to the particles populating cumulus clouds in the Earth's atmosphere. This was a theory that was developed by Dutch astronomers, mainly under the leadership of Professor H.C. van de Hulst. The idea was that cosmic dust grains continually condense out of the gas in space rather like droplets out of a saturated vapour. A fundamental difficulty with this idea that impressed us in the 1960s lay in the operation of the condensation process itself. Within interstellar clouds the density of gas atoms was so low that condensation centres, or nuclei, would not form at anything like an adequate rate. Thus the ice grain theory began to look somewhat shaky in the early 1960s.

In 1962 we proposed that the particles might be made of carbon in the form of graphite, the material from which the familiar 'lead' of pencils is made. In the mid 1960s the observation of stellar spectra at ultraviolet wavelengths became possible, and such observations soon appeared to bear out a prediction that we made for an absorption peak due to graphite particles. We thought at the time that graphite particles were formed in gases that escaped out into space from a group of stars known as the carbon stars, in the fashion of smoking chimneys. After a while, however, we realised that this idea, although initially promising, could not be the complete answer. We modified the model to include mixtures of ice and graphite, possibly ice condensed as mantles around graphite cores, and later to mixtures of particles made of graphite and rocky material. We had varying degrees of success with each of these models, models which might collectively be described by the term 'inorganic'.

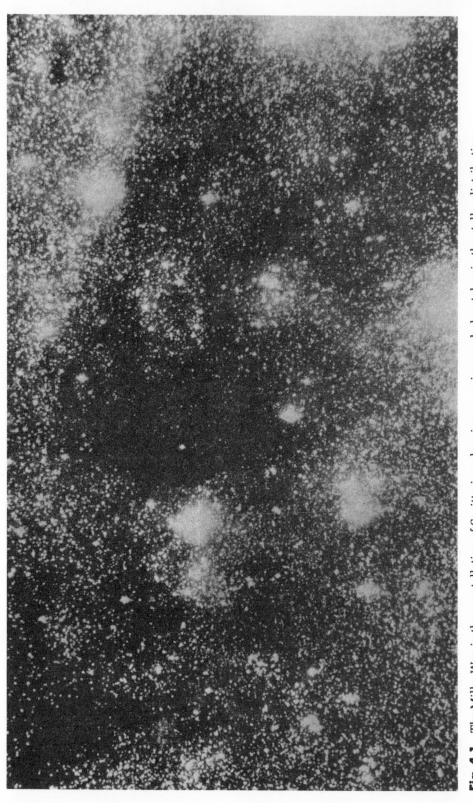

Fig. 4.1 The Milky Way in the constellation of Sagittarius, showing conspicuous dark patches in the stellar distribution, indicating the presence of obscuring dust clouds. (Courtesy of the Mt Wilson and Palomar Observatories)

By the mid-1970s the data that were available about the properties of cosmic dust were so extensive and precise that our failure to obtain good agreements with our inorganic dust models led to a measure of embarrassment. This failure was to us at least an indication that we were barking up the wrong tree. At long last it dawned on us that there must be a logical connection between the composition of the gaseous organic molecules which we knew to exist in the interstellar clouds and the cosmic dust particles themselves. The important question was the direction of the connection: did the dust come from the organic gas, or did the organic gaseous molecules derive from the dust grains? Our first thought was that the connection went in the former direction, namely from gas to dust. We explored the possibility that cosmic dust grains were organic polymers, first considering formaldehyde polymers as being derived from the ubiquitous molecule of formaldehyde, H_2CO, known to exist in the gaseous state. Next we considered the polymeric molecule cellulose, and thereafter more complex biological polymers. All these models gave rather better agreements with astronomical data than the inorganic models we had considered earlier. The trail was blazed in whichever direction the astronomical data pointed, without any notion of where we were heading.

The correspondences with astronomical data improved as we proceeded along the trail, but perfect agreements continued to elude us until one day in 1979 when we arrived logically at the startling conclusion that cosmic dust had to be freeze-dried bacteria. It took us little time to verify that bacteria are similar in size to cosmic dust grains. The dust grains are remarkably similar throughout space, a property of grains that had remained enigmatic, to say the least, until that time. Furthermore, dust particles in interstellar space were known to have an abnormally low value of the index of refraction (a measure of the extent to which they scatter light) compared with the properties of inorganic models. It turned out by a remarkable providence that bacteria have exactly the right properties in this regard. Under exceptionally dry conditions, as in the low pressure environment of interstellar space, the free water in a bacterium would evaporate out of the porous cell walls, leaving a particle with

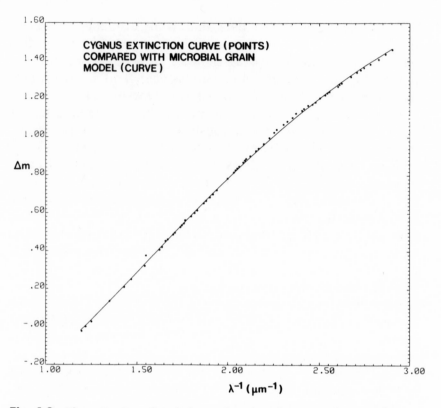

Fig. 4.2 The extinction of starlight as a function of wavenumber over the visual wavelength region (points) compared with the scattering behaviour predicted for dry bacteria (curve)

interior cavities. Such hollow particles behave as if they have a very low average index of refraction, a property which turned out to agree with the astronomical observations to a very precise degree, as seen in Fig. 4.2.

Further correspondences between the bacterial model and astronomical data unfolded in an impressive way as we began to probe this hypothesis in greater depth. Early in 1979 Shirwan Al-Mufti had begun a series of pioneering experiments to investigate the properties of bacteria in the laboratory under conditions that mimicked closely the conditions prevailing in interstellar space. Quite accidentally as a result of Al-Mufti's work we discovered a property that could be searched for on an astronomical scale, if bacterial particles were indeed widespread in space.

We found that over a certain infrared waveband, 2.9–3.9 micrometres, there was a distinctive signature of living material that might be looked for in sources of cosmic infrared radiation. This signature was closely the same for bacteria of all sorts as well as for more complex life-forms if they were powdered down to the sizes of bacteria. A crucially important cosmic source in this respect was one designated GC-IRS7, which had been observed by astronomers for some time, and which had already shown a hint of the feature we were looking for. This source is almost at the centre of our galaxy in the middle of the dark region shown in Fig. 4.1. The radiation emanating from this source not only traverses the dust in the local vicinity, but it passes through all the dusty material lying over tens of thousands of light years between the centre of the galaxy and the Earth.

Within months of completing our bacteriological measurements in Cardiff, two astronomers working in Australia, Dayal Wickramasinghe and David Allen, observed this same astronomical source again but with far higher spectral resolution than before, using the powerful Anglo-Australian Telescope at Sidings Spring Mountain in New South Wales. What emerged from this new observation surpassed all our hopes and expectations. The remarkably close agreement that showed up between the new data and the bacterial model of the dust is shown in Fig. 4.3. For us at least there was no need for further doubt, but as it happened there were other proofs that followed, confirming in our view the correctness of this model.

Another correspondence that impressed us tremendously came from the spectrum of cosmic dust within the Trapezium nebula in the constellation of Orion. Here dust particles were heated by surrounding stars to a temperature of about 175 degrees above absolute zero. The heated particles radiated their own energy over infrared wavelengths and in so doing displayed a characteristic spectral signature of the material of the dust in the wavelength range 8–12 micrometres. This spectral signature was once thought to support a model of the dust involving mineral grains, that is to say particles similar to household dust. But even the very best fits possible for such a model left a great deal to be desired, as can be seen in the upper panel of Fig.

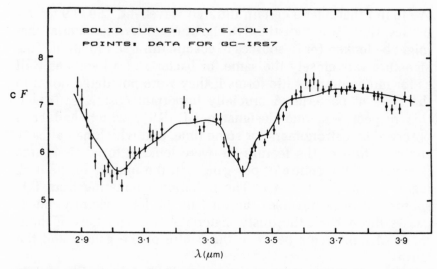

Fig. 4.3 The spectrum of infrared radiation observed by David Allen and Dayal Wickramasinghe for the galactic infrared source designated GC-IRS7 (points) compared with the predictions of the bacterial model (curve)

4.4. The lower panel shows what happens if the cosmic dust is of a biological nature. Here the curve was calculated from data obtained in the laboratory for a mixed culture of microorganisms actually present in a sample of water taken from the local River Taff, one of the many polluted waterways containing large microbial floras that criss-cross the British Isles. The river flora involved here included both purely carbonaceous microorganisms and numerous types of siliceous microorganisms (diatoms) where polycondensations of the molecule $Si(OH)_4$ occur in cell walls. This naturally occurring mixture of microorganisms might be thought to be fairly generally representative of the microorganisms that occur in the wider cosmos.

The upshot of these arguments and correspondences with data is that some form of microbiology must be operating on a cosmic scale. And indeed why not? To confine microbiology to the planet Earth, where perchance we happen to have stumbled upon it, is an artificial constraint, as unrealistic as it is restrictive. Nor is the Earth sealed away, microbiologically speaking, from the rest of the galaxy, either now or at any earlier epoch. Anyone concerned with sterilising surgical wards or spacecraft will tell

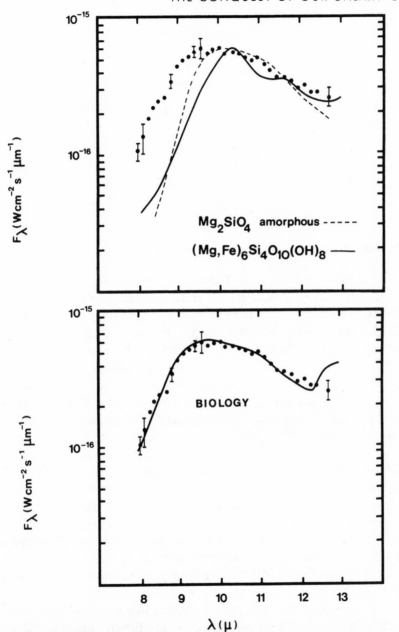

Fig. 4.4 The infrared spectrum of dust in the Trapezium nebula (points) compared with models of the dust (curves). The upper panel is a comparison with two types of mineral silicate particles, hydrated and amorphous ; the lower panel is a comparison with a naturally occurring mixture of microorganisms

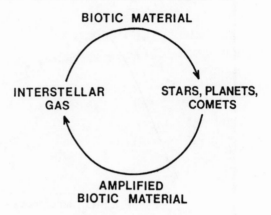

Fig. 4.5 The cosmic amplification cycle for life

you that the indomitable hardihood of microorganisms, combined with their readiness to multiply, presents a formidable problem. Given the right nutrients and conditions, a bacterium divides into two in about a couple of hours. These then further divide into four, eight ... and so on, until essentially all the nutrient medium is used up or until the culture becomes extremely concentrated. If a culture medium in which bacteria can grow can be imagined to extend from a terrestrial laboratory to reach out to the edge of the galaxy in a contiguous manner, the total conversion time of nutrient matter into bacteria would be scarcely a few weeks. This simple thought experiment drives home the point that cosmic biology cannot easily be subjugated. As far as conversion of nutrients to microorganisms goes, the actual conversion time will be longer only because the galactic culture medium is necessarily broken up into separate, discrete bodies within cometary clouds, as we have noted in earlier chapters. A realistic conversion time across the galaxy on such a picture is undoubtedly longer than weeks, but not much longer than a geological epoch consisting of 100 million years.

The cosmic cycle that serves to amplify microbiology in our galaxy, and indeed in all galaxies, is shown schematically in Fig. 4.5. The fundamental driving force in this loop is the explosive amplification of life in every galactic niche that develops, where living cells encounter conditions in which liquid water and the right inorganic nutrients come together. Apart from causing microorganisms to increase in their numbers, the cycle of Fig. 4.5 could also operate in a way that modulates astrophysical processes such as the formation of new stars. For example, the rate at which new stars form might be sensitively controlled so as to maintain at an appropriate rate the production of a component of bacteria that remains viable from one star-forming site to another. Too high a rate of star formation could break the cycle because of the potentially destructive effects of ultraviolet radiation from new-formed stars, while too small a rate would also tend to cut off the loop. In all, about 10^{10} cycles in this feedback loop would have been completed since our galaxy was born, one for every Sun-like star. No matter how few the starting number of viable bacterial cells might have been, the exponential nature of microbial growth which operates at each circuit of this cosmic cycle would ensure that the whole galaxy is eventually colonised by cosmic life.

These ideas can be seen to have an intimate connection with both the origin of life on Earth and with the origins of microorganisms in the Universe. It also links with the concept of panspermia discussed by Svante Arrhenius somewhat less than a century ago. Panspermia is of course a coined word meaning the dispersal of life on a cosmic scale. Arrhenius essentially pointed out that bacterial spores have sizes that are correct for being acted on by the pressure of starlight, in such a way that they can be propelled from one starsystem to another. The important dimension of a bacterium, which is about 0.3 micrometres, is just right for this purpose, as it is also right for explaining the scattering properties of starlight. These coincidences would be a trivial matter of chance if one excluded their greater role in the cosmic scheme of things.

Svante Arrhenius also correctly pointed out that bacterial spores have an almost incredible degree of robustness, in correspondence to conditions they would experience in space. He argued, for instance, that they could survive the low temperatures of interstellar space, this conclusion being reached after studying the known behaviour of bacteria and seeds under exceedingly low temperatures in the laboratory. Arrhenius was, however, unable to know the appropriate conditions with regard to X-rays and ultraviolet radiation which at first sight might be thought to be deleterious to the survival of life. We shall return to these matters later and show that an adequate measure of survival will always be ensured. For the moment let us note that the operation of the feedback loop in Fig. 4.5 does not require preservation of microbes in a viable condition from one cycle to another to any more than a fractional degree. Even if only a handful of cells are viable amongst all that reach a newly formed comet cloud, the cycle could proceed with positive feedback, easily dominating all inorganic processes that can also produce particles of similar sizes.

We now proceed to connect these ideas with another historical development of over a century ago. Louis Pasteur (1822–1895) was among the first scientists to contemplate the origin of microorganisms in an experimental way. Pasteur's monumental work of 1857 is generally thought to have laid the foundations of modern biology, as well as of both the germ theory of disease and of fermentation. But unwittingly perhaps he can also be seen as having paved the way for the cosmic theory of life. Even as late as the mid-nineteenth century it was widely believed that lowly life-forms such a bacteria could originate de novo from organic material during the fermentation of wine and the souring of milk – a view that seemed to imply the spontaneous generation of microorganisms. By a series of elegant experiments Pasteur showed that in reality each microorganism is derived from a pre-existing parent microorganism, a bacterial or fungal spore. In one famous experiment Pasteur placed sterilised nutrient broth in which microorganisms can grow in two separate flasks. In one flask he fitted a hooked pipe over the neck in such a way that bacterial spores from the air were prevented from reaching the culture broth. In the other flask the neck was left open to the air. The flask exposed to the air became turbid, showing microbial growth, whilst the other remained clear.

This experiment showed to most people's satisfaction that microorganisms are indeed derived from parent microbes, just as in the case of larger life-forms, plants and animals alike. Thus the ancient belief in spontaneous generation gave way to the new empirically tested fact that life is always derived from life – a fact that is seen to be as decisively valid at the present time as it is throughout the entire geological record. Life-forms existing today are connected at each generation by a causal thread to life-forms in earlier generations, on and on back in time to geological epochs as represented by fossils, all the way back to the earliest microbial fossils that we now know were deposited some 3.5–3.8 billion years ago.

This logical thread leads us inevitably to the all-important question of the origin of the first microorganisms on the Earth, a question that we have already touched on in earlier chapters. Spontaneous generation in a primordial soup in terrestrial oceans has been proposed, but for well over half a century this has remained a proposition that has defied proof. The basic idea of this proposition, which was initially discussed by A.I.Oparin and J.B.S.Haldane, is that a mixture of organic molecules, which is supposed to have developed in the Earth's oceans, began to undergo a sequence of chemical transformations that eventually led to life. It does not take much imagination to see that this was indeed a revival of the old doctrine of spontaneous generation in a new guise. Laboratory experiments that have been carried out with a view to establishing this theory have been singularly without success in producing structures with even the slightest resemblance to biological structures. The disproof of spontaneous generation arrived at empirically some 130 years ago remains as cogently valid today, and for the same fundamental reason: at the level of assembly of the molecular building blocks of life – the amino acids and the nucleotides – into informational structures (e.g. the enzymes), living systems are far too complicated for asssembly by random processes occurring in a terrestrial primordial broth.

There is of course no logical requirement for life to have started from scratch on the Earth. Our planet was assembled from cosmic material along with the Sun, the comet cloud and other planets some four and a half billion years ago. This cosmic material would necessarily have contained its fair share of microbial particles such as we found to exist in vast quantities everywhere in space. The entire planetary system is today surrounded by tens of billions of cometary objects, each equivalent in volume to an organic laboratory with a floor area of a hundred square kilometres and a height equal to the Empire State Building in New York. This immense system must have been a relic from the time when the outermost planets Uranus and Neptune condensed. Each one of the cometary bodies would have acted as amplifiers of cosmic biology on a very short timescale in the manner we have discussed earlier. Those cometary objects that did not become engulfed in the planets Uranus and Neptune, cometary objects a hundred billion strong, now surround the planetary system in the form of a gigantic spherical halo. It is from this halo of comets that individual members are perturbed by the gravitational action of passing stars in such way as to become new long-period comets. Comets carrying microorganisms must in our view have interacted with the Earth throughout its entire history, over a period of 4.5 billion years. Just as Pasteur had found that the growth of microorganisms under laboratory conditions is governed by the presence of pre-existing organisms, could not life on Earth originate by contamination from comets? To deny this logic would, we argue, be a travesty of all the available facts from astronomy and biology alike.

Looking outwards from the solar system we see ample evidence for the birth of new stars from clouds of gas and dust. The solar system was formed from a cloud of material very similar to compact clouds to be seen today in the Orion nebula. We know that new stars are being born almost continually from such clouds, the youngest stars in the Orion nebula having shone for less time than humans have walked on this planet. The precise details of the processes involved in the spawning of new stars are not yet understood, but we know that myriads of cosmic dust grains, which we have argued to be mostly bacteria, play a crucial role. They go into regions where stars form and emerge from the neighbourhood of newly formed stars in vastly amplified numbers. These observations are entirely consistent with our contention that we are witnessing in these regions the explosive amplification of biology within cometary objects around new stars. There is no satisfactory explanation for such phenomena within the framework of so-called conventional astronomy. We shall return to these matters in a later chapter.

We conclude the present chapter by discussing a crucial test for the correctness of this point of view. It is a necessary consequence of these ideas that bacteria must be space-hardy, and so they are found to be. A viable strain of the bacillus *Strepto-*

coccus mitis was recovered from a TV camera after two years of exposure to conditions on the surface of the Moon. Bacteria can be brought down to near zero pressure and temperature without loss of viability, provided suitable care is taken with the experimental conditions. Bacteria can survive after exposure to pressures as high as 10 tonnes per square centimetre, and after flash heating under dry conditions at temperatures up to 600 degrees Celsius. Viable bacteria have been recovered from the interior of an operating nuclear reactor. A fraction of bacteria remain viable even after extremely heavy flash doses of high energy radiation, upwards of a megarad, while it seems that bacteria can repair themselves continuously in an environment of high radiation intensity, to the extent of repairing tens of thousands of breaks in their nucleic acid structures. These are not properties one would have expected to evolve on the Earth, but they are all properties necessary for survival in space. Damage from interstellar ultraviolet radiation is no serious problem either, because such radiation is very easily shielded against. A thin skin of graphitised material around a single bacterium or clump of bacteria effectively shields the interior from destruction by ultraviolet light. Bacteria could not be better equipped for the conquest of galaxies.

FIVE

Interstellar communications

A primary thesis of this book is that cells, among which is a viable fraction, cells particularly in the form of freeze-dried bacteria, are prevalent everywhere in the galaxy. Any planet or planets that arise anywhere in the galaxy will therefore be in receipt of this cosmic bounty of life, and wherever the planetary conditions are favourable for survival such life could take root. With the billions of candidate stars that are available with conditions similar to the Sun, it does not stretch credulity too far to suppose that the survival and development of cosmic life on planets must be rather commonplace.

The question that concerns us in this chapter is not the survival of primitive life, but its elaboration into complex and sophisticated forms of intelligence. This surely happened on the Earth and must have resulted only from the assembly of genes from cosmic bacteria and other genetic structures such as viruses and viroids. We can see intermediate steps in this assembly process within the present-day cells of plants and animals. For instance the chloroplasts of plant cells are thought to be inclusions of photosynthetic algal-type cells, and parts of the DNA of higher animals like ourselves is known to comprise viral sequences, since from time to time replicating viruses break loose from our DNA, in a reversal, as we think, of the way our DNA was originally formed.

Evolution of life on the Earth can be viewed as a direct consequence of our continued exposure to cosmic genes, and had it not been for such exposure terrestrial life would not, as we see it, have proceeded beyond the stage of single-celled structures. On this picture similar evolutionary patterns and assembly operations must surely have taken place on countless other planets. If intelligence and scientific technology of the kind that evidently surrounds us is the end-product, or intermediate end-product, of this evolutionary assembly process one might wonder whether our entire galaxy, even the entire Universe, might not be teeming with intelligent life. To maintain otherwise might be seen to be scarcely conceivable if for no other reason than the staggering vastness of the cosmos.

The Sun, our parent star, is one of some 100 billion stars in the Milky Way, and our Milky Way system itself is one quite ordinary spiral galaxy amongst billions of other similar systems in the Universe. To considerations of vastness should be added also the likelihood of planets circling around stars. Present-day estimates suggest that such occurrences are highly probable, at least to the extent of providing perhaps a billion billion of planetary systems which are suitable homes for life throughout the observed region of the Universe. In our view the basic components of life, including those that contain the blueprint for intelligence, must have come to be shuffled into 'thinking' creatures on a vast number of other planets outside the Earth. A concept of intelligent life that is Earth-centred can thus be ruled out, we think, as a narrow untenable extension of pre-Copernican thought.

Experience has shown, however, that a purely rational attitude to this subject is often difficult to achieve. A geocentric, anthropocentric view of the world has been held with great tenacity by human beings in diverse cultures and from time immemorial. It seems almost part of human nature to adopt such an instinctive stance. Whilst rational attitudes have sometimes succeeded in overcoming prejudice, Earth-centred, human-centred ideas have always died hard. The geocentric model of the world still persisted even in Europe for more than a hundred years after Copernicus, when at last it was reluctantly replaced by the idea that the Earth goes around the Sun. The present-day resistance to accept the logic of cosmic life and of cosmic intelligence could well be a further extension of our predilection for a Universe centred upon ourselves.

Some scientists have recently asserted that intelligence compatible with either space travel or interstellar communication must be confined to the Earth because otherwise we would long since have been colonised by an alien civilisation. This argument is as naive as it is fallacious. Human technology of 1988 permits space travel, it is true, but only to a limited extent, and reasonably only within the confines of the solar system. Manned interstellar travel lies only within the realms of science fiction and will almost certainly remain there. It is not at all clear that further advancements in the field of space technology or further evolution of human life would lead to a capability, let alone a desire, for large-scale galactic travel. Half a century ago travel around the Earth was surely a romantic ideal for many people. Now travelling is getting to be a bore. In the decades ahead most people would probably prefer to conduct their work at home in front of a computer screen with the world literally brought to their finger tips. The efficient way of learning about distant places is by transmitting information and receiving it, not by the tedious process of travel.

We could attempt to listen-in for broadcasts and messages from external intelligences, and we could also beam out coded radio signals in the directions of prospective planets. Two-way communication may at first sight seem to be ruled out because of the great distances to be bridged. But eventually it is possible that we would get used to the idea of waiting a few decades for responses from nearby planetary systems. At what frequencies do we broadcast and listen and what messages do we send? It is of course impractical to listen-in or to broadcast signals at all frequencies and in all directions. The nature of the messages to be beamed must also be carefully thought out so as to be potentially intelligible to alien civilisations. They would obviously not understand a language such as English or French, but they would almost certainly understand statements expressed in numbers about the properties of our physical environment. So all in all a fair number of intelligent choices on our part would need to be made before we can seriously think of communicating with our peers abroad.

There are a few well-defined frequencies in the spectrum of radio waves that seem especially fundamental to us and possibly also to other civilisations as well. Neutral atomic hydrogen in space, for example, has a spectral line centred at 1420.40575 MHz, a frequency that is a factor 10 higher than the upper end of frequencies used for radio transmissions in the familiar FM waveband. Interstellar hydrogen atoms, the most abundant chemical species in the Universe, are known to radiate at this frequency. The hydrogen line is one that obviously suggests itself, and it seems reasonable to guess that intelligent beings with radiotelescopes elsewhere in the galaxy might choose to broadcast near this apparently fundamental cosmic frequency. So it appeared at any rate to G. Cocconi and P. Morrison in the year 1959. Their now classic paper blazed the trail for a serious pursuit of the problems of interstellar communications. Their essential idea was that if an alien intelligence has set out to make contact with the likes of ourselves, 'they' will endeavour to make it as easy as possible for us by transmitting at a frequency or frequencies that are obviously well-known.

In the year 1959 the hydrogen line frequency was the only one detected by astronomers, but since then the radio lines of over 60 atoms, molecules and radicals (fragments of molecules) have been discovered. The popular belief still is that the hydrogen line would be the frequency of choice. On the other hand, some alien civilisation could argue that the great abundance of radiation from hydrogen that occurs naturally in the galactic environment would lead to a high background level at this frequency, against which a coded signal would, in some situations at least, be lost or difficult to detect. One or more other frequencies that correspond to less abundant molecular species like the hydroxyl radical (a fragment of the water molecule) could in some ways be considered a better bet. Since life depends on the presence of water, this frequency might conceivably be thought to have an enhanced significance. Clearly a good deal of trial and error and experimentation would be needed before we hit on the 'correct' frequency, but this might not prove intractable, for the choices fortunately are not unlimited.

The current search for cosmic intelligence got off the ground following a historic petition written by the astronomer Carl

Sagan and countersigned by 72 international scientists, including 7 Nobel Laureates. The petition read as follows:

The human species is now able to communicate with other civilisations in space, if such exist. Using current radioastronomical technology, it is possible for us to receive signals from civilisations no more advanced than we are over a distance of at least many thousands of light-years. The cost of a systematic international research effort, using existing radio telescopes, is as low as a few million dollars per year for one or two decades. The program would be more than a million times more thorough than all previous searches, by all nations, put together. The results – whether positive or negative – would have profound implications for our view of our universe and ourselves.

We believe such a coordinated search program is well justified on its scientific merits. It will also have important subsidiary benefits for radioastronomy in general. It is a scientific activity that seems likely to gather substantial public support. In addition, because of the growing problem of radio frequency interference by civilian and military transmitters, the search program will become more difficult the longer we wait. This is the time to begin.

It has been suggested that the apparent absence of a major reworking of the galaxy by very advanced beings, or the apparent absence of extraterrestrial colonists in the solar system, demonstrates that there are no extraterrestrial intelligent beings anywhere. At the very least this argument depends on a major extrapolation from the circumstances on Earth, here and now. The radio search, on the other hand, assumes nothing about other civilisations that has not transpired in ours.

The undersigned are scientists from a variety of disciplines and nations who have considered the problem of extraterrestrial intelligence – some of us for more than 20 years. We represent a wide variety of opinion on the abundance of extraterrestrials, on the ease of establishing contact, and on the validity of arguments of the sort summarised in the first sentence of the previous paragraph.

But we are unanimous in our conviction that the only significant test of the existence of extraterrestrial intelligence is an experimental one. No *a priori* arguments on this subject can be compelling or should be used as a substitute for an observational program. We urge the organisation of a coordinated, worldwide, and systematic search for extraterrestrial intelligence.

Thus it was that the now famous SETI (Search for Extra Terrestrial Intelligence) programme was conceived and born. Compared with other activities connected with Space Research SETI was cheap, as the petition writers had stressed. The partial deployment of already existing resources in the form of radio telescopes and radioastronomers was all that was initially required.

Long before this modern petition and independently of the ideas of Cocconi and Morrison, the radioastronomer Frank Drake used the 85 ft Tatel radio telescope at the National Radio Astronomy Observatory at Green Bank, West Virginia, for precisely the same goal. After several false alarms when Drake apparently picked up signals from nearby trucks and airplanes which at first were mistaken for cosmic signals, he began his search for extraterrestrial intelligence in earnest in the spring of 1960. His initial attention was directed to two nearby stars, Epsilon Eridani (10.7 light years) and Tau Ceti (11.9 light years), in a programme which came to be code named OZMA, after the Princess in the Land of Oz in the famous children's stories by Baum. Once again Drake relates a gripping story concerning his 'biggest and best false alarm'. One day in 1960 Drake turned his radio telescope antenna toward the star Epsilon Eridani. A rhythmic pulse was detected loud and clear, ticking away at the rate of 8 pulses per second. Drake was exhilarated to think that he might have found 'them'. When the antenna was moved away from the direction of the star the signal subsided exactly as it would do if the source was really localised around Epsilon Eridani. But when the telescope was turned again to the star the signal had gone. So far the hypothesis was not destroyed. Drake could still have maintained that ET had broadcast a message that lasted for only a couple of minutes. But disproof came when the same signal was detected later by a receiver that was non-directional. These signals are thought to have arisen, most probably, from one of those ever present airplanes in the sky.

After this catalogue of disappointments in the 1960s the SETI began to take off slowly in the succeeding decades. Starting in 1973 two American astronomers, B.Zuckerman and P.Palmer, turned their radiotelescopes in the directions of some 500 nearby stars with spectral types similar to the Sun and used modern high-speed computers to store, analyse and search their data for hidden signals. This search still continues and others have followed in a similar course.

What sort of signals should we be looking for? Any regular rhythmic signals of the type seen by Drake in his false alarms could be potential pointers to intelligent communication. Of course nearby Earthly sources have to be ruled out and a source that remains fixed relative to the stars must be identified. Only then can we be sure that the signal has a cosmic origin. As for the nature of the signal itself we cannot do better than quote from the pioneering article by Cocconi and Morrison:

> We expect that the signal will be pulse modulated with a speed not very fast or very slow compared to a second, on grounds of band-width and of rotations. A message is likely to continue for a time measured in years, since no answer can return in any event for some ten years. It will then repeat from the beginning. Possibly it will contain types of signals alternating throughout the years. For indisputable identification as an artificial signal, one sequence might contain, for example, a sequence of small prime numbers of pulses, or even simple arithmetic sums.

To give an idea of what type of message we might hope to receive from space, and how we might decode it, Drake presented his colleagues at a conference in Green Bank with a sheet containing a sequence of 551 ones and zeros (see Fig. 5.1). The challenge was to find out what type of message was implied, if this had indeed been received from an alien planet. In the event only one conference participant is reported to have successfully decoded Drake's message. The result is shown as the picture in Fig. 5.2, in which certain vital statistics of human existence are revealed. The way this is decoded is as follows. First note

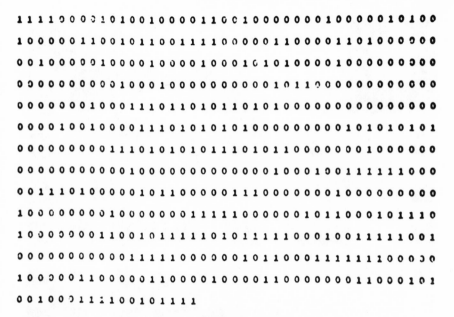

1 1 1 1 0 0 0 0 1 0 1 0 0 1 0 0 0 0 1 1 0 0 1 0 0 0 0 0 0 1 0 0 0 0 0 1 0 1 0 0
1 0 0 0 0 0 1 1 0 0 1 0 1 1 0 0 1 1 1 1 0 0 0 0 0 1 1 0 0 0 0 1 1 0 1 0 0 0 0 0 0
0 0 1 0 0 0 0 0 1 0 0 0 0 1 0 0 0 0 1 0 0 0 1 0 1 0 1 0 0 0 0 1 0 0 0 0 0 0 0 0 0
0 0 0 0 0 0 0 0 0 1 0 0 0 1 0 0 0 0 0 0 0 0 0 0 1 0 1 1 0 0 0 0 0 0 0 0 0 0 0 0
0 0 0 0 0 0 1 0 0 0 1 1 1 0 1 1 0 1 0 1 1 0 1 0 1 0 0 0 0 0 0 0 0 0 0 0 0 0 0 0
0 0 0 0 1 0 0 1 0 0 0 0 1 1 1 0 1 0 1 0 1 0 1 0 0 0 0 0 0 0 0 1 0 1 0 1 0 1 0 1
0 0 0 0 0 0 0 0 0 1 1 1 0 1 0 1 0 1 0 1 1 1 0 1 0 1 1 0 0 0 0 0 0 1 0 0 0 0 0 0
0 0 0 0 0 0 0 0 0 0 1 0 0 0 0 0 0 0 0 0 0 0 0 1 0 0 0 1 0 0 1 1 1 1 1 1 0 0 0
0 0 1 1 1 0 1 0 0 0 0 0 1 0 1 1 0 0 0 0 0 1 1 1 0 0 0 0 0 0 1 0 0 0 0 0 0 0 0 0
1 0 0 0 0 0 0 0 1 0 0 0 0 0 0 0 1 1 1 1 1 0 0 0 0 0 0 1 0 1 1 0 0 0 1 0 1 1 1 0
1 0 0 0 0 0 0 0 1 1 0 0 1 0 1 1 1 1 1 0 1 0 1 1 1 1 0 0 0 1 0 0 1 1 1 1 1 0 0 1
0 0 0 0 0 0 0 0 0 0 1 1 1 1 1 0 0 0 0 0 0 1 0 1 1 0 0 0 1 1 1 1 1 1 0 0 0 0 0
1 0 0 0 0 0 1 1 0 0 0 0 0 1 1 0 0 0 0 1 0 0 0 0 1 1 0 0 0 0 0 0 0 1 1 0 0 0 1 0 1
0 0 1 0 0 0 1 1 1 1 0 0 1 0 1 0 1 1 1 1

Fig. 5.1 Frank Drake's original message for transmission to a prospective
extraterrestrial intelligence given as a sequence of 29 × 19 ones and zeros

that there are 551 ones and zeros in all and an arithmetician
would soon spot that this number is the product of the two
prime numbers 29 and 19. This suggests a 29 × 19 matrix format,
pointing to the use of square paper with a 29 × 19 grid. Starting
from the top line and working downwards, translate 1 into a
black square, 0 into a white square and the result is the picture
in Fig. 5.2. Whether extraterrestrial intelligent beings would
think in this way is another matter, but at least it is logical
and it is a possibility. Following the same line of logic we may
think of dispatching similar pictures into space, in the hope
that they would eventually be received by a civilisation at some
distant place in the galaxy.

In 1974 Drake, while he was Director of the Ionospheric
Observatory at Arecibo, Puerto Rico, had the opportunity of
broadcasting such a message as part of a ceremony to inaugurate
the Observatory's refurbished 305 metre dish. The Arecibo message in ones and zeros is shown in Fig. 5.3 and its pictorial transla-
tion is shown in Fig. 5.4. This picture in the form of ones and
zeros was beamed at the hydrogen frequency in the direction

Fig. 5.2 Message in Fig. 5.1 translated into pictorial form

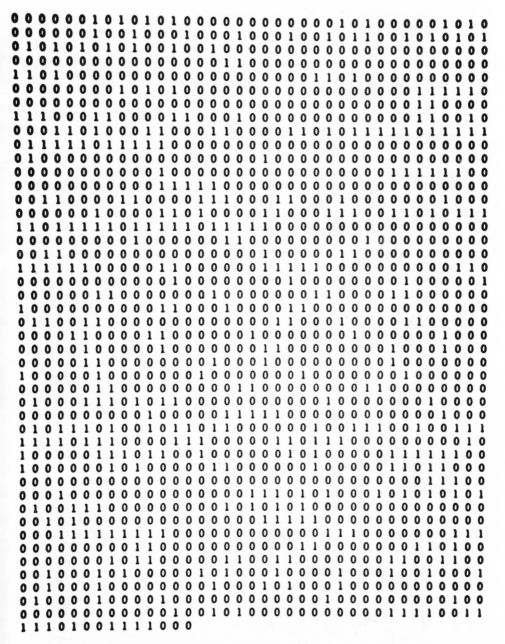

Fig. 5.3 A message actually transmitted from Arecibo in 1974 directed to the globular cluster M13 in the constellation of Hercules some 25,000 light years away

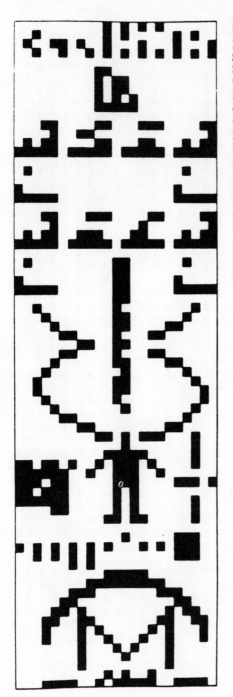

Fig. 5.4 The Arecibo message in picture form. The uppermost band gives the numbers 2–11 in binary form ; next the atomic numbers of the key elements for life – hydrogen, carbon, nitrogen, oxygen, phosphorus ; the next three bands give chemical formulae for the components of DNA ; the broad middle bands show the structure of the DNA double helix ; then the current human population, the shape of a human, and the height of man compared with the radio message wavelength ; next is a band showing the structure of the solar system ; finally the diameter of the Arecibo radiotelescope compared to the radio wavelength

of the globular cluster designated M13 in the constellation of Hercules. The choice of direction for beaming was dictated by the consideration that the presence of some 300,000 stars closely packed within M13 would multiply the chances that the signal will actually be picked up in the future and decoded by an extra-terrestrial civilisation associated with one of the 300,000 stars. Of course one has to be sure that the signal when it reaches its destination has not been so attenuated as to be indistinguishable from background radio noise. Fortunately the radio background of normal stars is weak, and it is not too difficult to beam signals that stand out against such a background. The globular cluster M13 is some 25,000 light years away. When Drake's message reaches the cluster about 25,000 years hence, and if an ET there happens to be observing the Sun at the 1420 MHz frequency at this date, its radio brightness will be enhanced within the signal pulses above the solar background by a factor of about 10 million. Such signals would be hard to miss!

A limitation in most of the early programmes connected with SETI arose from the need to tune a radiotelescope to a single channel. Although in order to get good signal-to-noise quality one requires individual radio channels to have very narrow bandwidths, such a requirement turns out to be highly restrictive. It does not allow for slight shifts about any magic rest frequency, such as the hydrogen line, that would inevitably occur due to relative motion between observer and source and for a variety of other reasons. If the transmitting waveband is narrow, as it should be, the location of a slightly displaced frequency is not as simple as tuning a radio knob! The old techniques used by Drake in the OZMA programme permitted only a few channels to be explored, more or less in a random way. This limitation is currently being overcome by a pioneering programme of research spearheaded by Paul Horowitz of Harvard University. Horowitz and his colleagues have devised techniques that permit 'listening in' to 65,000 narrowband channels essentially at the same time, and this number is expected by further refinements to be increased to 8 million in the early 1990s. The probability of detecting extraterrestrial intelligence is thus coming to be increased by factors that are truly astronomical. The full impact of this emergent technology for the SETI programmes has not yet been fully grasped by the scientific community, but already the new devices have been successfully tested at the Arecibo radiotelescope. Searches are under way with a vengeance, most of these being sensibly directed at specified targets including some 773 F,G and K Sun-like stars out to a distance of 81 light years. The frequencies being searched are generally confined around the H-line and the OH-line (the water hole) for reasons we have already discussed.

In addition to searches for articulated messages put out by external civilisations, why not also look for 'incidental' signs of their existence? The activities of high-tech or super high-tech societies would surely show up in a variety of unintentional ways. Thus astronomers R.Freitas and F.Valdex recently thought fit to use the Hat Creek radiotelescope to search for the 1516 MHz line of the hydrogen isotope tritium in the neighbourhood of normal stars, on the grounds that this nuclide has a short half-life of 12.5 years. If this unstable element is indeed found around a star similar to the Sun it could only imply an artificial origin, as for instance in a nuclear reactor. Another idea has been to search for the electromagnetic 'garbage' that would inevitably be generated by external civilisations as an incidental by-product of their existence. To test this proposition W.T.Sullivan and S.H.Knowles used the Arecibo telescope once again to look at the Moon for reflections of television and radio signals emanating primarily from our own planet. They found clear evidence of the stronger TV stations around the world, and of course even clearer evidence of military radar signals. It was inferred as a result of this experiment that the space surveillance radar operated by the National Aeronautics and Space Administration would be clearly detectable by an extraterrestrial civilisation possessing an Arecibo-type technology at a distance of some 20 light years. This technique, politely referred to as 'eavesdropping', has great possibilities that would need to be exploited in the future. At the moment such searches have been confined to only a few nearby stars.

Another important contribution to the discussion of extraterrestrial intelligence was made some years ago by physicist Freeman J.Dyson. He pointed out that the eventual advancement of a human-type civilisation in a solar-type planetary system would be in the direction of utilising all the energy output of our parent star and all the material resources that are reasonably accessible. To this end he suggested that we contemplate an advanced human-type technology capable of disassembling the entire mass of a Jupiter-type planet to form a spherical screen around the Sun. In this way it might be possible to utilise, not merely the solar energy falling on our planet, but all the energy of sunlight trapped within such a shell. The material available for human technology and human expansion is also not necessarily confined to what is there in the terrestrial biosphere, but could be expanded to include the mass of the giant planets. In this way the accessible material for intelligent life-forms to increase their numbers and their activities would be multiplied by a factor of 10,000,000,000 and the available energy would be multiplied by 10,000,000,000,000. If such an advanced civilisation occurs cosily tucked in around a distant solar-type star, all that would be observable to us would be garbage photons at infrared wavelengths. Sunlight at visible wavelengths would be taken up and effectively degraded into the infrared. It might then be difficult to distinguish between such an object and a newly formed star (a protostar) surrounded in its cocoon of dust. A possible discriminant could be the location of such an object in relation to nearby stars, or to other similar objects. A source of thermal infrared radiation found in isolation from other similar objects may be thought to be a strong candidate for intelligence, whereas sources that occur within a well-recognised association of young stars are almost certainly embryonic stars or stars about to be born.

SIX

The viral vector

The inheritable information in all of us, the blueprint for life, is carried in our genes that lie at the heart of cells. Genes are made of the famous double-helix molecule DNA (deoxyribonucleic acid), within which the arrangement of the four units, guanine, adenosine, thymine and cytosine, in sequences that sometimes run into tens of thousands, code for proteins, and in turn for all the functions of our cells. It is the precise order within these sequences that distinguishes between the multitude of life-forms that inhabit our planet. A curious property of our DNA that has only recently come to light is that over 90 per cent of it is normally inert – it is not used for making proteins but is merely copied in our genes, repeatedly, from cell to cell, from generation to generation. In certain relatively rare diseases, viral particles are seen to emerge from this normally inert DNA, thus suggesting that the whole of our DNA might be derived from viruses. The point of view that we shall develop in this chapter is that this DNA, the blueprint for life, was put together from genetic fragments that actually came from outside the Earth.

In addition to the double-stranded nucleic acid, DNA, there is also the single-stranded form, RNA or ribonucleic acid, which carries essentially the same type of information as DNA, and RNA and DNA are known to be convertible from one form to another within cells. Thus DNA and RNA are essentially equivalent as far as genetic information is concerned.

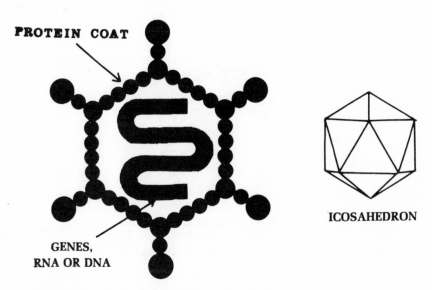

PROTEIN COAT

ICOSAHEDRON

GENES,
RNA OR DNA

Fig. 6.1 Schematic depiction of a virus exhibiting an icosahedron structure

A typical virus is a tiny particle measuring about one ten thousandth or less of the size of a pinhead. It is made up of nucleic acid, either RNA or DNA, and is surrounded by a double-layered shell of proteinaceous material. The external shapes are highly regular geometrical figures, a common shape being the icosahedron, a solid which has 20 triangular facets. The structure of such a virus particle is shown schematically in Fig. 6.1. The spikes protruding from the corners are extensions of the protein shell that actually help the virus to recognise suitable hosts and eventually to penetrate a host cell. There is an intimate cell-virus relationship depicted in Fig. 6.2, a relationship that is almost conspiratorial.

The virus attaches itself to specific sites on the surface of a host cell. Then it is quickly engulfed by the cell's outer membrane and is effectively sucked into the interior of the cell. Next, the host cell proceeds to strip the virus of its outer protein coat, and thereafter it takes its instructions from the invading virus. The instruction is 'stop what you are doing and produce more viruses like me'. This is instantly obeyed by the cell. Finally, the newly formed viral particles burst forth from the cell wall, almost invariably destroying the cell in the process,

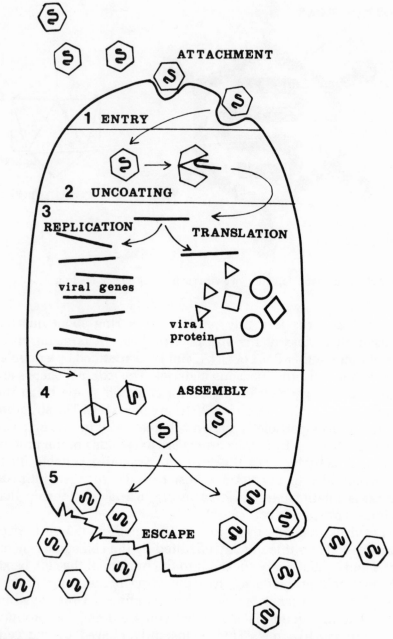

Fig. 6.2 Schematic depiction of a cell-virus interaction from entry of viral particles at the top to replication, and finally to egress of newly formed virus particles at the bottom

and causing other cells to become infected by the viral particles so released.

Viruses are in general quite fussy about the kind of cells they attack. A particular virus – e.g. the influenza virus – selects a species or a small range of species, and also a subset of cells within a species. This property is widely cited as an argument against an extraterrestrial origin of viruses. You could ask the question: how could a virus that evolved in some place other than the Earth be able to replicate in terrestrial host cells? In other words, how could a virus coming from outside the Earth know ahead of its coming here the nature of the cell which it is going to find? Our answer is simple. It is true that the incoming virus cannot know in advance what host cell it is going to encounter. But we, the host cells, could certainly know and recognise the virus, if our genes contain viruses of a similar kind. According to our point of view we must have had a long and continuous interaction with viruses stretching back over billions of years. So a virus such as the influenza virus that comes in today (or tomorrow) simply seeks out those species in whose ancestral lines certain aspects of this same virus were incorporated perhaps millions of years ago.

On the basis of the standard earth-bound theory of life, this apparent conspiracy on the part of cells of higher life-forms to admit viruses is baffling. If there were no favourable positive aspect to the process of viral infection, it is surprising that evolution could not have conferred total protection from viral attack. In the many steps of the virus-cell interaction a blockage of access at any one step could have been easily arranged, one would have thought. And the logical possibility of preventing the overwriting of the cell's genetic programme must surely exist, for the much larger information content of the host cell could very easily contrive to swamp the trivial information content of a virus. If viruses had no positive role to play it is hard to think that such a defence was not developed over long evolutionary timescales in creatures that are as highly evolved as we seem to be. The reason that this has not happened must be, we think, that the entry of viruses into our genetic material was a prime prerequisite for evolution.

This view is supported by considering the specificity of viruses to particular species in a little more detail. Large hospitals routinely culture specifically human viruses but they do not use human cells in the process. Cells of quite different species are used, sometimes as widely diverse from human cells as the cells of a chicken. The curious point therefore arises that viruses are not specific when they multiply within individual cells. It is their attack on whole animals that is specific. One wonders what the difference is. The answer is surely the operation of the immunity system of the whole animal. The specificity of viruses in most cases is to immunity systems not to cells. This suggests an entire inversion of our usual way of looking at viral diseases. Instead of thinking of viruses outwitting cells, which is not really plausible because of the exceeding paucity of the genetic material in many viruses, we should think of viruses being deliberately invited into our cells. We should think of our immunity system as constantly scanning the newcomers, with a view to permitting our genetic system to seize hold of any that might be valuable from an evolutionary point of view. Those that are clearly useless are nobbled without ceremony. Those with possible potential are encouraged to interact with the cells, with different animals scanning different viruses according to their separate needs. Only if a virus is close to being useful to us is it permitted to attack.

This view goes against the grain since we inevitably tend to think of viruses as bad because we suffer individually from them. However, the suffering of the individual is irrelevant to evolutionary biology. What matters is the occasional success, not a million failures. If this were not the correct picture it is otherwise difficult to understand why the body produces its best defence, the substance interferon, only last in the case of serious viral attack. If the body were really seeking to dispose of viruses, one would expect interferon to be produced first, not last.

In our view the first life-bearing comet to arrive at the Earth and to find our planet congenial to life was one that brought living cells some 3800 million years ago. What seem to be primitive life-forms with the shape of bacteria as well as yeast-type cells have been found fossilised in early sedimentary rocks. The arrival of primitive life-forms could not have stopped at that distant time, however. Bacteria, larger cells and fragments of cells and viruses and viroids have on our point of view continued to arrive through cometary injections right up to the present time. It has been this steady arrival of genetic material that has led to a progressive evolution of terrestrial life. With each arriving comet, viral particles (probably derived from cells) covering an astronomical wide range of properties could interact with life-forms already present on the Earth. Some viral genes simply add on, others have the effect of attenuating life-forms that become, as we say, 'diseased'. No amount of shuffling of the genetic material of a primordial bacterium could by itself lead to a flowering plant on the one hand, or a man on the other. Evolution we do not deny, but as we see it, evolution could only have been driven from outside. The viral vector from space has a crucial role to play.

The available evidence from the fossil record bears ample testimony to this point of view. The sudden surges in evolution and diversification of species of plants and animals, and the equally sudden extinctions evident in the fossil record, point to sporadic additions of cometary genes with the arrival of major new crops of comets. Cometary genes could become grafted onto pre-existing biological stock leading to dramatically new lines. Contemporaneously the effects of epidemic disease could lead to extinctions in relatively short timescales.

A dramatic example is provided by the extinction of the dinosaurs 65 million years ago, a subject we have already touched upon in an earlier chapter. These highly successful reptiles that strutted our planet for over 100 million years disappeared in a very short period of geological time. We know now that a comet or comets were somehow involved because of the unusually high content of the element iridium that was recently discovered in the deposits where the dinosaurs were found. We do not, ourselves, think this enormous extinction event could have been caused by a purely physical process such as the spraying of cometary dust around the Earth or the actual physical impact on the Earth of a large object. Fig. 6.3 shows the distribution of various genera of small sea-living creatures and of plankton on either side of the 65 million year boundary. While some lines are seen to stop abruptly, others are seen to start equally sharply at this same boundary, at the precise moment of geologi-

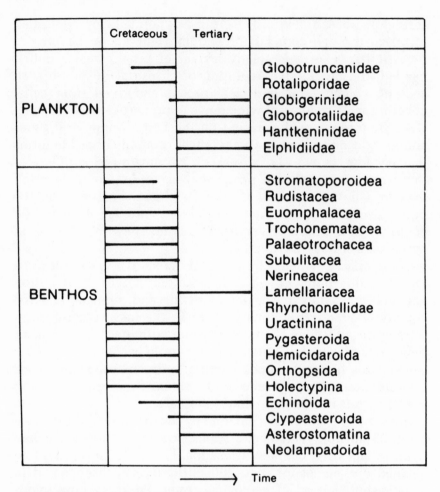

Fig. 6.3 The bars indicate periods of existence of various genera of small sea-living animals, with plankton from the surface water and benthic creatures from considerable oceanic depths. (Adapted from D.A.Russell in *Syllogeus*, no. 12, National Museums of Canada, 1976)

cal time when the dinosaurs became extinct. We recall also that these effects were by no means confined to small marine creatures and plankton. Nearly half the genera of all animals were lost at the same time.

Another important event that took place at about this time was the great surge of mammals that was to lead in due course to the emergence of Man. Although some evidence of mammalian

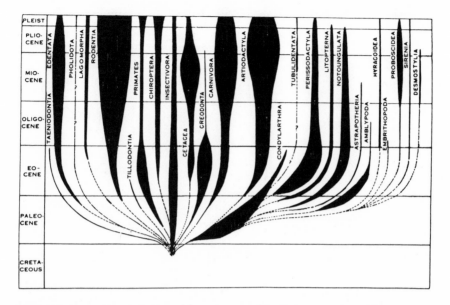

Fig. 6.4 The fossil record for the orders of mammals is shown by the solid areas, the dotted continuations being only conjectural. The widths of the solid bands indicate variable numbers of species in the different orders

characteristics can be found at earlier times, the main expansion of mammalian orders occurred 65 million years ago. The solid bands of Fig. 6.4 show the fossil record for the orders of mammals, with the dotted extensions drawn to correspond to the way evolution is thought to have occurred according to conventional theory. The striking aspects of Fig. 6.4 are first that no detailed connections have been found between the orders, and second that the conjectured connections all converge to the same point in time. Both the extinctions of some orders and the emergence of others on a massive scale are in our view an indication of a major genetic storm that came with the arrival of cometary viruses 65 million years ago.

Looking across the fossil record in its entirety one sees that this event is by no means unique. The fossil record can be seen as largely static with no evolutionary changes occurring except at several sharply defined moments in geological time. As with the dinosaur extinction event, such changes as occur are not confined to individual genera or small groups of related genera, as one might expect on the usual 'Darwinian' picture, but they

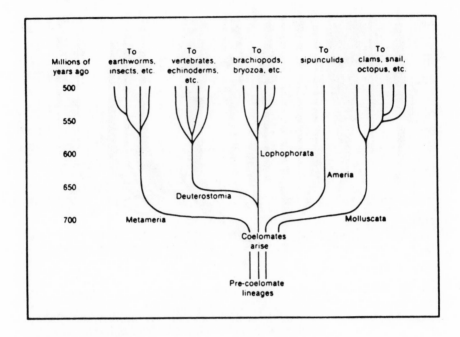

Fig. 6.5 Radiations from a primitive stock of coelomates

extend over a broad sweep of phyla and orders all at once. Evolutionary biologists have described these events as implying an evolution by 'punctuated equilibrium', although this nomenclature itself offers no explanation whatsoever of the cause of the so-called 'punctuation'. Fig. 6.5 shows the apparent radiations from a primitive stock of coelomates with major branchings and splittings of evolutionary lines occurring, all at once, some 570 million years ago. A similar situation shows up in depictions of the diversification of mammalian orders where the emergence of new orders takes place at quite definite moments in geological time. The same is also true of plant life where the branch called angiosperms led to the sudden emergence of flowering plants 130 million years ago. The same is true for the origin of eyes, insect wings and so on and on in an almost endless catalogue. Information for brand new developments comes 'out of the blue'.

On the standard Darwinian picture such phenomena as these make little or no sense. It is believed by Darwinists that the full spectrum of life as we see it today, as well as in the past, is accounted for by the steady accumulation of copying errors in genetic material from generation to generation. It is stated according to the theory that the accumulation of copying errors, sorted out by the processes of natural selection, the survival of the fittest, could account for both the rich diversity of life and for the steady upward progression from a bacterium to Man. This view not only goes against present-day knowledge of copying error rates, but it should, if it were correct, lead to smooth evolution, not to sudden major jumps.

In our book *Evolution from Space* and in other works we have argued strongly against this proposition. We agree that sequential copying would accumulate errors, but on the average such errors must lead to a steady degradation of information, at any rate in simple systems that lack bisexual protection. It is absurd in the extreme to suppose that the information provided by one single initial cell can be upgraded by copying to produce a man, and all other living things that inhabit our planet. This standard view in biology is similar to the proposition that the first page of Genesis copied billions upon billions of times would eventually accumulate enough copying errors, and hence enough variety and diversity, to produce not merely the entire Bible, but all the holdings of all the major libraries of the world! It can actually be proved mathematically that a uniform line of some initial cell (a clone) must necessarily become degraded as time goes on, the reverse of what classical biology maintained. The processes of mutations, gene doublings and natural selection can only produce very minor effects in life as a kind of fine tuning of the whole evolutionary process. There is above all an absolute need for a continual addition of information for life, an addition that extended in time throughout the entire period of the geological record.

In our view every crucial new inheritable property that appears in the course of evolution must have been added to a pre-existing stock in the form of cometary genetic fragments. Although ape and Man admittedly have much in common biochemically, anatomically and physiologically, they are at the same time a world apart. It is hard to accept that the genes for producing great works of art or literature or music, or for developing skills in mathematics, emerged from chance mutations of genes present in our ancestral apes long ahead of their having any conceivable relevance for survival in a Darwinian sense. We shall return to discuss these matters later, but for the time being let us note that, as is the case with the most primitive life on our planet, all such changes and discontinuities had to be implanted from outside.

It is a consequence of this point of view that any life-form that was somehow shut away from external viruses would cease to evolve. There is some evidence to suggest that such a fate may have come to many species of insects and crustaceans. Not long ago a colleague who is well versed in the study of bees informed us that the oldest fossils of bees, beautifully preserved in an almost mummified form in tarry deposits, revealed a creature 26 million years old and yet identical in every detail to modern bees. Although a confirmed Darwinist by training and choice, this somewhat bewildered professor admitted that his discovery had considerably shaken his faith. The only explanation he could offer was that the bee was 'perfectly adapted' right from the outset and that the steps leading to this highly adapted form were conveniently lost from the fossil record. The same type of story runs through much of biology. An expert on crustaceans will tell you that the shrimp has remained virtually unchanged from the beginning, and other examples are not lacking in all those instances where the fossil record happens to be reasonably complete. Our attempts to learn about the susceptibility of insects and crustaceans to viral attacks have met with little success so far. Perhaps reliable information that relates to this matter does not exist. But at any rate such viral attacks do not appear to be conspicuous phenomena in these creatures at the present time. If they are indeed attacked by viruses we would have to predict that they have at least succeeded in devising a mechanism to block the assimilation of cosmic viral genes into their own DNA.

The general reaction that greets the proposition that bacteria and viruses may be arriving from space is one of disbelief, a disbelief that stems essentially from educational prejudice to which we are all subject. We are taught from an early age that life arose on the Earth in some sort of primordial soup, and if this were indeed true terrestrial life may seem to be quite reasonably earthbound. This remarkable proposition is taught not merely as just one possibility amongst others, but as a verified, proven fact. Yet there was never any positive evidence for it, and nowadays, as we have seen, there is a good deal of evidence that goes against it.

Setting prejudice aside for the moment, let us note that this planet of ours is unquestionably an open system; it is not hermetically sealed or isolated from the rest of the Universe. On the contrary there is direct evidence of a continual injection of cometary material to the Earth. Photographs of the Earth taken recently from satellites at certain ultraviolet wavelengths have actually shown the arrival of small comets at a steady rate. The rate of input of cometary material is not known with any degree of certainty, but it is thought to be in the range from 1000 to 10,000 tonnes per year. If even a minute fraction, say .001 per cent, were in the form of viruses the annual intake of viruses would add up to some 10^{23}. In addition to viruses the cometary input would also contain bacteria, perhaps numbering 10^{21} per year, which is about the same as the number of raindrops which fall to the ground all over the Earth in the course of a year.

It has been repeatedly stated by critics without proof that cometary microorganisms would all be destroyed by heating as they plunge into the Earth's atmosphere. This can be shown to be untrue. We have carried out laboratory experiments on the survivability of bacteria with respect to flash heating on atmospheric re-entry and found that heating even to a 1000 degree temperature above absolute zero for a few seconds under dry conditions does not lead to any loss of viability. It is true that spacecraft re-entering the atmosphere would be heated to the point of sterilisation at its surface, and certain types of cosmic particles, e.g. meteoroids of sizes of the order of a millimetre, are destroyed by frictional heating. But this phenomenon is sensitively dependent on the size, composition and the degree of fluffiness of the incoming particles. We have repeatedly argued in our writings that clumps of bacteria, individual bacteria, as well as viral-sized particles must all survive atmospheric entry to a significant degree. Survival is also ensured for even the most delicate biological structures embedded within loosely compacted cometary fragments that come to be dispersed only within the stratosphere, or even lower down in the atmosphere. In the latter case the deposition of biological material could be highly localised on the surface of the Earth. This could be relevant to the occurrence of highly localised outbreaks of bacterial and viral diseases as we shall discuss in the next chapter.

Cometary microorganisms reaching the upper atmosphere – say to a height of 100 km – begin by falling under gravity but they are quickly sifted out according to size. Particles of bacterial size continue to fall under gravity and could reach ground level in a matter of a year or two. Viral-sized particles become trapped at a height of some 20–30 km in a stratospheric 'trap' and further descent is largely controlled by global mixing circuits of stratospheric air. These circuits have an essential seasonal character, bringing down common viruses to ground level in seasonal cycles.

If large numbers of bacteria and viruses are being constantly added to the Earth, the reader might wonder whether such incidence could not be proved by direct detections in the atmosphere. Detection experiments of this kind are not as easy as one may think, however. A major difficulty arises in separating a true extraterrestrial flux from a population of bacteria of terrestrial origin that might be carried up into the atmosphere in high winds, or carried up with the experimental equipment itself. If, though, one flies sterilised balloon packages to sufficient heights, these difficulties could be in a considerable measure overcome. Since vertical movements of the air are in general extremely feeble in the region of the stratosphere, one would not expect particles of bacterial sizes to be carried in air currents to heights much above 20–30 km. Any biological particles discovered with equipment that is initially sterilised at greater heights would have a high probability of coming from space.

A series of balloon flights into the stratosphere, reaching heights above 40 km, was made by American scientists in the 1960s. The results were all positive in a way that baffled the investigators concerned. Viable bacteria were recovered that could be cultured by standard techniques. The equipment was said to be sterilised before each flight, and two identical instrument packages were flown, one of which was exposed to the atmosphere, while the other was not. The unexposed package served as a control. Since bacterial cultures were not recovered from the control package any possibility of a laboratory contamination can be ruled out.

These early experiments gave results ranging from 0.1 to 0.01 cells per cubic metre in the stratosphere, with a density actually increasing as heights were increased from the range 60–90,000 feet to 120,000 feet. This is the opposite to the trend one would expect for bacteria blown upwards from the ground.

In the late 1970s Russian experiments of a similar kind sought to collect air samples from higher up in the atmosphere, in the mesosphere, above a height of 50 km. Rockets fired into the high atmosphere expelled detection equipment attached to a parachute device. Film was exposed over various height ranges, with particles collected on the film being sealed as the equipment descended out of the height range in question. Recovered film was then examined in the laboratory for microorganisms. After three such flights some 30 cultures were grown of bacteria obtained from heights of 50 km to 75 km. The evidence from both the American and Russian experiments would thus seem to favour the hypothesis of bacteria incident from space. If, however, one is impelled through prejudice to regard this proposition as intrinsically implausible, there would be no difficulty in brushing aside the unpalatable results on the grounds of 'possible contamination'.

As far as we know, there have so far been no searches for viral particles in the upper atmosphere. Such detections could well prove more difficult because of the requirement of collecting large enough numbers of virus to obtain cultures by the techniques that are currently deployed, e.g. the injection into chick embryos. The choice of a prospective host cell system could also prove problematical for unknown viral types that might be recovered.

Collections of particulate material in the lower atmosphere, at heights below 25 km, have consistently turned up populations of particles that resemble bacteria and viruses to varying degrees. The Australian physicist E.K.Bigg has for many years been recovering particles that are similar in exterior characteristics to microorganisms.

The most likely route to ground level for an extraterrestrial microorganism that comes to be dispersed in the stratosphere is via the rain. The micro-oganism would effectively serve as a nucleus around which a particle of water ice could grow. For many years scientists have been baffled by the problem of how clouds come to be seeded so as to produce rain. An atmospheric cloud of saturated water vapor at a temperature of 0°C or slightly lower does not spontaneously turn into rain without either the formation within or the introduction from outside of 'freezing nuclei'. The problem was highlighted repeatedly in unsuccessful attempts over the years to seed rain clouds using inorganic crystals.

Over three decades ago the distinguished Australian physicist E.G.Bowen discovered a remarkable connection between 'freezing nuclei' in rain clouds and extraterrestrial particles. He showed that there was a link between the frequency of freezing nuclei detected within clouds and the occurrence of meteor showers. Meteor showers occur at regular times in the year as the Earth crosses the trails of debris evaporated from short-period comets. Although larger particles that enter in this fashion would be evaporated quite high in the atmosphere, microorganisms could survive and so be able to act as freezing nuclei.

E.K.Bigg described the following event that has a relevance in the present context:

> In 1968 I obtained by an extraordinary coincidence a
> photograph from an altitude of 30 km of a meteor entering
> the atmosphere almost tangentially 2 hours after sunrise.
> From this photograph it is evident that a great deal of volatile
> material evaporated from the meteor at heights not less than
> 270 km when it would have been only mildly warm.
> Microorganisms contained within the icy matrix would
> presumably have had a good chance of being dispersed at that
> time....

Fig. 6.6 shows a frequency distribution of the daily detections of freezing nuclei in the month of January for the three years 1954, 1955, 1956. As Bowen has pointed out, the peaks at January 13, 22, 30 for freezing nuclei also coincide with peaks in the rainfall curves at many stations throughout the world over very

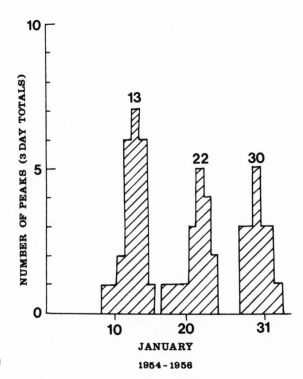

Fig. 6.6 Histogram of the dates in January when the concentration of freezing nuclei was observed to be high in 1954, 1955 and 1956. (Data from E.G.Bowen)

much longer periods of time. Also, quite remarkably, these three dates are nearly four weeks after the Earth crossed the meteor streams Geminids (Dec. 13), Ursids (Dec. 22) and the Quadrantids (Jan. 3). Similar correspondences with a 4-week lag time have been shown for all the major rainfall peaks averaged over many years and in many places. Bowen was simply reporting what he observed without any reference to a theory such as ours. Writing in *Nature* (177, 1121, 1956) Bowen reported as follows:

The hypothesis has therefore been advanced that dust from meteor streams falls into the cloud systems of the lower atmosphere, nucleates them and causes exceptionally heavy falls of rain thirty days after the dust first entered the atmosphere.

There are several aspects of this hypothesis which are difficult to accept, and many questions arise. Why should meteoritic dust be so active as a nucleating agent? Why does the time of fall appear to be so consistent from one shower to the next?

To answer the second question first, this apparent problem for inorganic meteoric dust ceases to exist if the icy meteoroids of the type seen by Bigg are aggregates of bacteria. The second question would require the bacteria to be all of essentially the same size. The 4 weeks would simply define the time of fall for particles of a particular size.

Bowen's second difficulty has been resolved by the most remarkable discoveries by Kolf Jayaweera and Patrick Flanagan which were recently published in *Geophysical Journal Letters* vol. 9, 94 (1982). Jayaweera and Flanagan collected air samples at several heights up to 7 km above the Arctic Ocean and found 10 types of bacteria and 31 different fungal spores which could lower the freezing temperature of water and serve as freezing nuclei. Although the height of 7 km does not automatically rule out contamination from the Earth for microorganisms, there are evident reasons now to connect Bowen's freezing nuclei with bacterial particles. Jayaweera and Flanagan write as follows:

> The species variety and numbers of microbial propagules found aloft exceed considerably those observed on the surface and in the soil near the Arctic Ocean.

It was found that a maximum concentration of about 10 microbial cells per litre was present within clouds and about 1 per litre in air outside clouds. It is also quite remarkable that the ice-nucleating properties of these bacteria as well as their sizes altered dramatically when they were cultured in the laboratory over extended periods of time. Over a ten-month run of a culture the authors claim that the activity of bacteria as ice-nuclei diminished significantly whilst their average lengths increased from 0.2 micron to 1.1 microns.

These developments are bizarre, to say the least, except on the basis of the theory we have proposed, the curious changes over a ten-month run occurring as the bacteria adapted themselves to the terrestrial environment which they had not experienced before.

SEVEN

Life-force and disease

If comets brought the first life onto our planet, it is a process that did not stop there. According to the ideas discussed in the previous chapter the influx of cometary microorganisms must have continued unabated to the present day, some simply adding to the microbial flora of our planet, others on occasion causing epidemic diseases in plants and animals.

The input of disease-causing bacteria and viruses is a proposition that has the advantage of being susceptible to proof or disproof. The detection of space-incident microorganisms would be considerably more difficult than detecting them at sites where small populations of incoming microbes of a particular type might be greatly amplified. Rather as physicists use amplifying detectors to observe small fluxes of incoming cosmic ray particles, so plants and animals can be regarded as detectors of pathogenic bacteria and viruses from space.

An incoming pathogenic microorganism from space can follow one of two logical paths. Either it could come from space and establish a reservoir in some group of humans, plants or animals, thereafter proceeding to propagate by case-to-case infection, or it may have the property of not being able to form a stable reservoir. In the first category, of which the viruses causing smallpox and AIDS are examples, epidemics occur when the reservoir essentially breaks its banks and spills over into the susceptible population. In the latter category, however, every attack, every new epidemic has to be driven directly from space. To identify a disease that comes in this second category is highly significant from the point of view of the thesis of this book. We shall argue now that influenza and many other acute upper respiratory tract infections appear to come in this latter category of direct space incidence.

First let us begin with influenza where, as we shall show, there are some rather decisive clues concerning the continuing incidence of viruses from space. Perhaps the most disastrous worldwide epidemic caused by this virus in recent times occurred in 1918/19 and caused some 30 million deaths. After carefully reviewing all the available information about the spread of influenza during this epidemic Dr Louis Weinstein wrote the following comment a few years ago:

> Although person-to-person spread occurred in local areas, the disease appeared on the same day in widely separated parts of the world on the one hand, but on the other took days to weeks to spread relatively short distances. It was detected in Boston and Bombay on the same day, but took three weeks before it reached New York City, despite the fact that there was considerable travel between the two cities. It was present for the first time at Joliet in the State of Illinois four weeks after it was first detected in Chicago, the distance between those areas being only 38 miles . . . .

Thirty years on it was the same story all over again. The worldwide epidemic of 1948 first appeared in Sardinia. A Sardinian doctor, Professor Margrassi, commenting on this epidemic writes thus:

> We were able to verify the appearance of influenza in shepherds who were living for a long time alone, in open country, far from any inhabited centre; this occurred almost contemporaneously with the appearance of influenza in the nearest inhabited centres.

In all the early accounts which one reads about influenza there is not the slightest hint that person-to-person spread played any important role in its transmission. The distinguished epidemiologist Charles Creighton maintained, as late as the final decade of the nineteenth century, that influenza is not a transmissible disease. In his book *History of Epidemics in Britain* (Cambridge University Press, 1891), he discussed the influenza epidemics of 1833, 1837 and 1847, in which medical opinion held that populations living over considerable areas were affected almost simultaneously. This evidence suggested to Creighton a 'miasma' over the land rather than a disease which must spread itself from person to person. If one substitutes for 'miasma' the phrase 'viruses from space', one has a position similar to that we have reached in the present book.

One of the most striking features of this whole story is that the technology of human travel has had no effect whatsoever on the way that influenza spreads. If influenza is indeed spread by contact between people, one would expect the advent of air travel to have heralded great changes in the way the disease spreads across the world. Yet the spread of influenza in the 1918/19 outbreak, before air travel, was no slower and no different in its detailed pattern of spread from what happened in recent times.

It is often said that clear evidence for the person-to-person transmission of influenza can be found in epidemic outbreaks within special communities – military bases, schools, isolated island people. However, it often happens in these data that the number of victims rises very steeply, doubling about every day, much more frequently than the 2–3 days incubation period of the disease itself. (Numbers on successive days might look like 3, 7, 20, 70, 240.) Such very steep rises are possible on the person-to-person infection theory only if each victim passes the disease to several other persons, the disease cycle being strongly supercritical. It is then a requirement that the outbreak cannot burn itself out until essentially all susceptible persons in the group have succumbed to it. Yet such steeply rising epidemic outbreaks often end with only some 25 per cent of the group affected, whereas antibody titres for the virus in question suggest that the susceptible percentage is much larger than this.

A remarkable opportunity for testing our theory arose nearly a decade ago in the winter of 1977/78 when an influenza subtype that had essentially become extinct some 20 years earlier came to the fore and began to affect children and young adults who had not encountered the virus previously. We studied the way

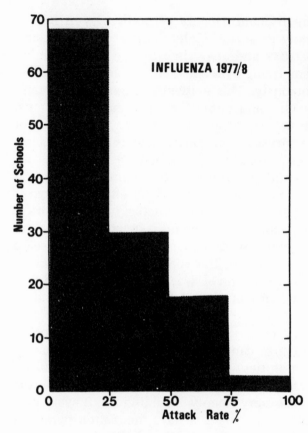

Fig. 7.1 Histogram showing the distribution of influenza attack rates in different schools in England and Wales during the 1977/78 epidemic

in which this particular outbreak showed up amongst school children in England and Wales, thereby regarding such children as amplifying detectors of the virus. Our analysis of the extensive data that we collected showed conclusively that the disease was not infective from person to person. It showed a very patchy incidence from school to school, and even from house to house within a single boarding school, fitting very well the patterns of incidence to be expected for a virus falling through a turbulent atmosphere, not the predictions for a horizontal infection model. Our survey involved a total of about 25,000 pupils, with a number of victims estimated to be some 8800, corresponding to an average attack rate of about 35 per cent. Fig. 7.1 shows the distribution of attack rates amongst the many schools that were involved. The person-to-person infection theory would have predicted a high peak for large values of the attack rate with a tailing off towards smaller attack rates. In fact what is observed is sharply opposite to this expectation. Only three out of well over a hundred schools at the extreme end of Fig. 7.1 showed the very high attack rates that have been claimed to be the norm.

All the diagnoses involved here were made by medical staffs in advance of our inquiries. Possibly other respiratory infections became associated with influenza in the diagnoses, but since January and February 1978 were months when the new brand of influenza was known to be prevalent, and since children of school age had no immunity against this virus, the bulk of the reported cases were most probably influenza.

The schools in our survey were all fee-paying, all with boarders sleeping together in dormitories. The degrees of association between pupils in dormitories, classrooms and at mealtimes could not have been much different from one school to another. If the virus responsible for the 8800 cases had been passed freely from pupil to pupil, much more uniformity of behaviour would have been expected. Already in Fig. 7.1 we see evidence of great diversity, with a strong hint that the attack rate experienced by a particular school (or house within a school) depended in a very detailed way on where it was located in relation to a general infall pattern.

The histograms in Fig. 7.2 shows the house-to-house attack rates for boarding houses (designated A, B, C etc.) in four independent schools. The Y-axis in each case gives the fluctuations above and below the mean attack rates for the separate schools expressed in units of the sample standard deviation. If influenza was indeed as infectious as it is claimed to be, such enormous fluctuations about the mean (six standard deviations in one case) cannot be explained. The virus was clearly air-borne, with incidence at ground level connected with meteorological factors and with the vagaries of aerosol transport through the lower atmosphere.

Another curious fact about influenza is that epidemics occur with a distinct seasonality in widely separated parts of the world. Cities on the same latitude belt show very similar patterns of incidence. This is seen in the data for the two cities

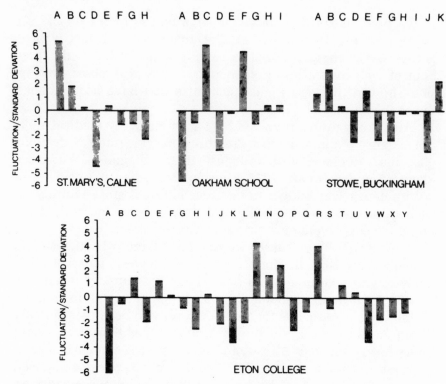

Fig. 7.2 Histograms showing fluctuations of influenza attack rates, house by house, in 4 schools during the 1977/78 epidemic

of Cirencester and Prague (Fig. 7.3) collected by Dr Edgar Hope-Simpson. A seasonal incidence pattern is also shown clearly in Fig. 7.4, where we have recently collected data from three countries: Sweden in the northern hemisphere, Sri Lanka on the equator and Australia in the southern hemisphere. These data represent month by month averages of incidence of influenza cases for approximately a decade in each country. With several new types of flu virus that occurred globally over the years in question it is mind-boggling on the horizontal person-to-person theory to understand how the disease was contained within the confines of seasonal winter peaks, separated by 6 months in the two hemispheres, despite the traffic of airline passengers travelling back and forth between Sweden and Australia. Surely the first plane load of Swedes arriving in Melbourne

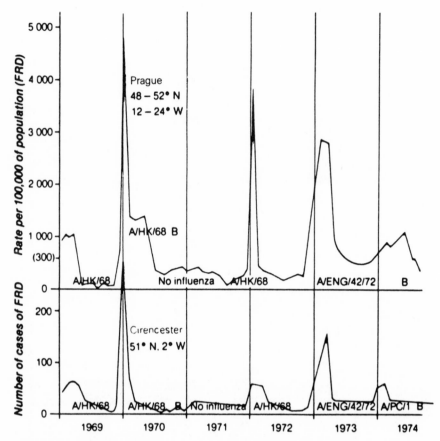

Fig. 7.3 Incidence of influenza in two cities, Prague (Czechoslovakia) and Cirencester (England), at roughly the same geographical latitude

even with one single passenger stricken with a new winter strain of Swedish flu, a brand new subtype say, would have triggered a major summer epidemic in Australia, if influenza is indeed person-to-person infective. The inference must be that it is not.

The pattern of incidence of influenza over all observable length scales – from tens of metres, as in the case of school houses, to thousands of kilometres across continents – is fully consistent with the transport of a viral-sized aerosol deposited periodically in the upper atmosphere. Major genetic shifts of the influenza virus, which occur about once every decade, must in our picture be connected with a new injection of cometary material. Periodic

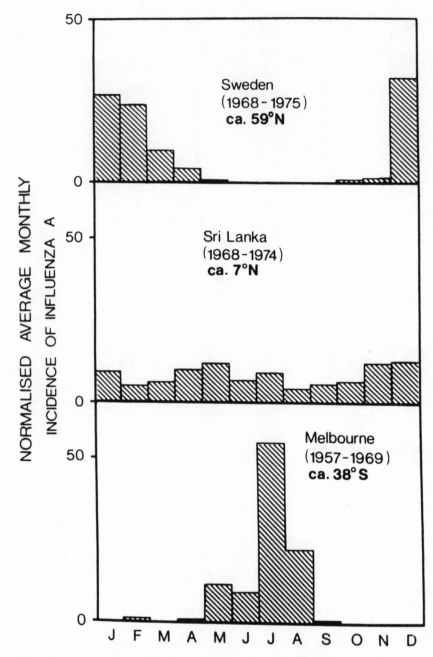

Fig. 7.4 Normalised average monthly incidence of influenza A in Sweden, Sri Lanka and Australia

injections could take place from different comets or from different parts of the same comet. In addition to dramatic genetic shifts that occur occasionally, there is also an appreciable genetic drift of the virus that occurs more or less continuously. This latter process, we think, is no more than a sieving out, according to the development of immunity within a human (or animal) population, a sorting out that must continually occur from an initially wide spectrum of viral genes introduced to the Earth at each cometary injection.

The first descent to ground level of a new crop of virus particles would occur eastward of the Himalayas for the reason that this lofty mountain range effectively punches a hole through the stratosphere, thereby providing a rapid drainage route for newly added virus. Such an effect is often seen for new shifts in the influenza virus, where populations in China and Hong Kong appear the first to succumb. Fallout elsewhere on the Earth subsequently follows the seasonal mixing cycle of the atmosphere.

Besides influenza there are also other viral and bacterial diseases that have all the appearance of being seeded from outside. For epidemics of the common cold, which occur in distinct waves across large tracts of country such as in the USA, all attempts to prove transmissibility under controlled conditions have led to failure. Experiments at the Common Cold Research Unit on Salisbury Plain, for instance, have not really provided convincing clues as to how this viral disease propagates as explosively as it does.

If viruses are pulled down seasonally from the upper atmosphere by meteorological effects, we would expect to find coincidences in timing of two or more different viral attacks. Fig. 7.5 shows the close similarity in the epidemic curves for two quite distinct viruses – the RS virus and the influenza virus – showing just the expected result.

For whooping cough or pertussis the situation is again quite revealing. This is a bacillus-caused childhood disease that has been well recognised since about AD 1578. It has been known for a while that epidemics of whooping cough occur in approximately 3.5 year periods. The conventional wisdom has been that this cyclic behaviour is due to the lag time required for the build-up of a new generation of susceptible children following

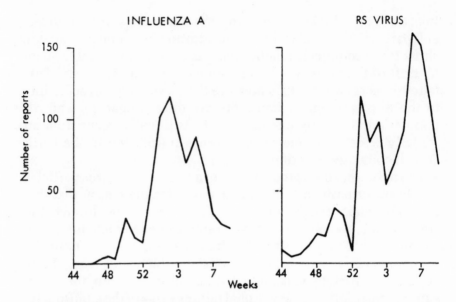

Fig. 7.5 Comparison of influenza A and RS virus incidence patterns during the winter of 1980/81. (Data supplied by the Communicable Disease Surveillance Centre, London)

an epidemic of the disease. An epidemic outbreak is thought to reduce the population density of susceptibles to such a low level that transmission through person-to-person contact is effectively stopped. The disease then smoulders until the susceptible population density builds up again from new births and migrations to a critical value – presumably every 3.5 years.

An interesting development for testing this theory arose because the population density of children susceptible to whooping cough was controlled during the period 1960–1975 using a vaccine that conferred active immunity against the disease. During these years the susceptible population fell to below 10 per cent of what it had been before. If the immunisation had been complete, then no epidemics would have occurred. In the real situation, however, where the immunisation was not 100 per cent, according to the usual theory epidemics should now have been more widely spaced in time. Fig. 7.6 shows the notifications of pertussis in England and Wales from 1940 to 1982. The 3.5 year cycle is seen to continue unabated throughout, except that fewer cases occurred in each epidemic during

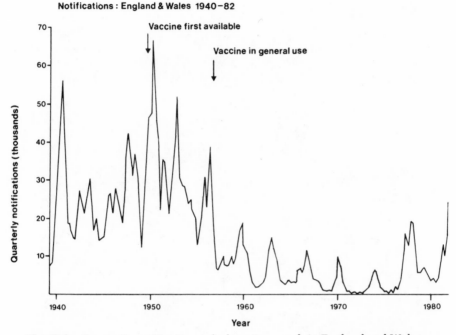

Fig. 7.6 Quarterly notifications of whooping cough in England and Wales, 1940–1982

the period 1960–1975. We note also that the decline of popularity of the whooping cough vaccine, which occurred in the late 1970s, has again done nothing to alter the nearly clock-work periodicity of the disease. This behaviour is impossible to reconcile with the conventional theories. It is also very hard to understand why the 3.5-year average period persists not merely in England and Wales, but throughout the world in countries with widely ranging demographic patterns, including the USA, Canada, Australia and India. Also puzzling is the fact that the same average period seems to have persisted as far as the record exists for over 150 years.

The periodicity of Fig. 7.6 is a strong indication that the supply of the pertussis bacterium is topped up every 3.5 years or so from some cometary source. A comet which suggests itself rather strongly is Encke's comet, which has an orbital period of 3.3 years. A photograph of the culprit comet is shown in Fig. 7.7.

It is certain that small particles emitted from Comet Encke

Fig. 7.7 Photograph of Encke's comet. (Courtesy of Yerkes Observatory)

actually enter the Earth's atmosphere. There are particles with sizes of the order of 1 mm, visible as meteors, forming meteor showers, the Beta Taurids from 23 June to 7 July and the Taurids from 20 October to 25 November, these being the times in the year when the Earth crosses streams of particles from Comet Encke. Since the Earth misses the path of Comet Encke itself, only particles ejected from the comet at explosive speeds can reach the Earth, which is to say only particles that have had their orbits appreciably changed by the ejection process can reach the Earth. In particular, it is essential that particles should go at ejection into orbits with much smaller inclinations to the Earth's orbit than the 12 degree inclination of the comet itself, a condition that requires the direction of explosive ejection to be nearly perpendicular to the plane of the cometary orbit.

The relevance of these considerations is that particles reaching the Earth from Comet Encke tend to have more energy per unit mass than the material of the comet, and so take a bit longer to go round the Sun. The orbital period of Comet Encke is 3.3 years, but the interval between successive arrivals of pertussis bacilli would be lengthened to almost exactly the value of 3.5 years as shown by the epidemic data.

If viral and bacterial diseases were propagated from person to person according to the commonly held view, then people living in high-density city areas should be significantly more at risk than those living in sparsely populated areas. From normalised attack rates plotted as a function of population density it would be possible therefore to prove the correctness or otherwise of this point of view. The circumstance that such data do not appear to exist, despite the cogency they would have, is interesting psychologically. Whereas people are avid to collect the slightest scraps of information that support conformist opinion, they are unremitting in their determination not to collect, or even to notice when collected, data which prove the opposite. It really needs no more than the absence of this simple but critical information to see that the commonly held view must be wrong.

One can say in general terms that if any major discontinuity existed between town and country, the population at large would easily be aware of it. Attack rates in overcrowded modern cities would be discernibly higher than in the country. Yet no such effect has ever been emphasised or reported. On a more quantitative level, Fig. 7.8 (see p. 104) shows data collected by Dr P.Jenkins, the Community Health Officer for the City of Cardiff. It gives figures for three diseases of infective jaundice (a viral disease), whooping cough and measles, obtained quarterly from the heavily populated Cardiff city area (after normalising to 100,000 population) and from the Vale of Glamorgan, much of which is very rural. Thus each disease in each quarter of a period of 5 years yields a point in Fig. 7.8, which has been plotted on two scales, one suited to quarters when numbers were comparatively small and the other suited to epidemic situations. The regression line drawn is not far from a 45 degree slope, showing at most only a slight bias, and moreover a bias that goes the wrong way for horizontal transmission. It is the lightly popu-

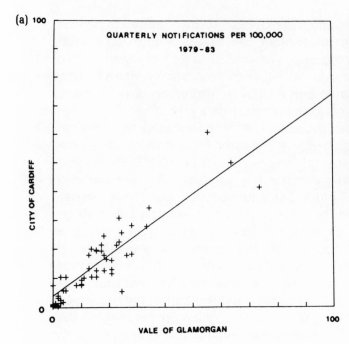

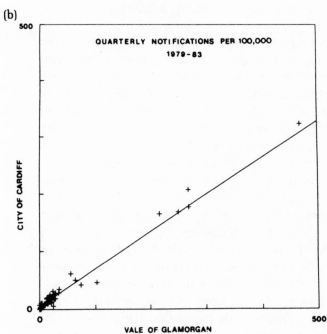

Fig. 7.8 Correlation between quarterly notifications of whooping cough, measles and infective jaundice in the City of Cardiff compared with the Vale of Glamorgan. Each point refers to a particular disease for a particular quarter in the period 1979–1983. Fig. 7.8a is plotted on a scale to show pattern for relatively low values of attack rate, Fig. 7.8b includes data for those quarters where higher attack rates were found

lated Vale of Glamorgan that on a normalised basis appears slightly worse affected. In other words one is marginally more at risk if one lives in the sparsely populated rural district than in the city! On our point of view the incidence at ground level of all air-borne bacterial and viral diseases will be determined by convection patterns of the air, and it would be entirely reasonable to find that the heat of cities produces an upwelling of air which in turn would somewhat lower attack rates in the cities.

There is scarcely a limit to the evidence one could adduce both from present-day medical records and from history. Historically, many bacterial and viral diseases have a record of abrupt entrances, exits and re-entrances onto our planet, exactly as though the Earth was being seeded at periodic intervals. In the case of smallpox, the time interval between successive entrances appears to have been about 700–800 years. Since man is the only host of the smallpox virus, global remissions of this disease lasting for many hundreds of years are very hard – almost impossible – to understand according to usual ideas. From the conventional point of view one has to say that the virus disappeared for a long period and then somehow reappeared in its original form, a requirement that is hard to understand.

There are many puzzles too in the medical annals of more recent times – puzzles that are resolved if we accept that pathogenic bacteria and viruses are falling from the skies. For instance, there was a group of about 500 Trio Amerindians who until quite recently had lived in the dense forests of Surinam, out of contact with the rest of humanity. When the forests were cleared and the tribe was discovered by anthropologists, it was found that there were polio victims amongst them who seem to have caught poliomyelitis at times roughly coincident with epidemics in cities hundreds of miles away. There is no conceivable way by which the forest-dwelling Surinam Indians could have caught polio from city dwellers. But the city dwellers and the Indians could have both contracted the disease if the causative viral particles rained on them from above.

The body of evidence that supports our contention is truly overwhelming. The urgency to come to grips with this situation and to recognise the space incidence of disease is highlighted by present-day efforts to control such diseases as Legionnaires' disease, meningitis and AIDS. Whilst the relatively harmless influenza virus, a regular visitor from space, is introduced to the Earth high in the atmosphere and is dispersed near ground level in evaporating rain drops that are inhaled, other routes of entry would apply for rarer diseases. A small comet disintegrating low in the atmosphere could lead to pathogens being brought down in rainstorms that are geographically highly localised. The pathogens could either be inhaled directly as for respiratory viruses, or they could be introduced into local water supplies from which they are later dispersed.

The comets responsible for a new disease such as AIDS are likely to be relatively rare objects. This virus was unknown in the human population a dozen years ago. Its sudden injection into the human population over the past ten years appears to point to a source that is external to the Earth. The virus that causes AIDS (Acquired Immune Deficiency Syndrome) was first detected in human populations of central Africa. A group of green monkeys native to the African continent has also been found to carry a closely related virus and the supposition has been made that the virus was in some way transferred from monkeys to humans. The much greater probability, in our view, is for both humans and green monkeys to be more or less simultaneously infected by a cometary viral pathogen. As in the case of influenza, the primary genetic sequences in the virus injected from space would have possessed a considerable degree of intrinsic spread, with the result that different AIDS viruses could have been selected by different species. A recent discovery of a feline AIDS virus in a cat in San Francisco also points to the occurrence of independent introductions amongst several species, say over a dozen or so years. A cat in San Francisco is hardly likely to have been infected by the African green monkey!

It seems most likely that the primary entry of the AIDS virus came from infected rainwater entering through skin lesions, whether in monkey, human or cat. It is probably no coincidence that the first entry into the human population occurred in a Third World country where footwear is hardly used, and skin abrasions are consequently common. Subsequent transmission, however, has proceeded through human contact, with each person passing the virus to at least one other person on the average.

In human societies it would be necessary for the virus to enter a social group that is sexually promiscuous or polygamous, if it is to turn into epidemic proportions. As things stand at the moment the threat of at least a partial extinction of the human species is a stark possibility that might have to be faced. The virus enters essentially through the human reproductive cycle and strikes at the very heart of our immune system, our capacity to combat disease.

A disease that continues to take a significant toll of lives is viral meningitis. Clusters of exceptionally high incidence are seen to persist in certain areas for several years, but the disease also shows a distinct seasonal preference for the winter months. Recent studies have revealed that a high proportion of the normal population are carriers of the implicated organism, N.Meningitidis, and this fact seems to us to indicate that the disease must require an additional causative agent. The winter effect, the space clustering and the exceedingly low infectivity seen for this disease in our view all point to a virus, a virus that has not so far been identified.

For Legionnaire's disease the involvement of a bacterium that is very commonly found in ground and surface water (from which domestic water supplies are derived) is also somewhat of a puzzle. Hospital cooling towers that are often blamed may well have only an incidental connection, merely serving to multiply and disperse the organism. In an outbreak of Legionnaires' disease that occurred in 1985 in the Staffordshire district, and in an earlier outbreak in Washington DC, heavy rainstorm conditions preceded the onset, pointing to a route connected with contaminated water supplies. The nearly coincident boundaries of the recent outbreaks of meningitis and Legionnaires' disease in the Stroud area of Gloucestershire also point in the same direction by suggesting a common environmental factor. A major international effort to carry out a rigorous and continuous microbiological surveillance of rainwater and groundwater on a worldwide scale seems to be long overdue.

EIGHT

The fabric of the world

Human beings living in the latter part of the twentieth century have witnessed feats of intellect and imagination that might in earlier generations have been thought unattainable. Scarcely a century after the dawn of the industrial revolution Man has probed both the minute world of the atom and the vast external cosmos with almost comparable dexterity, revealing a Universe rich in wonderment. In this chapter we trace the developments in both atomic physics and astronomy in broad terms, to set the physical background against which the 'cosmic life-force' must operate.

Probing the minute began in earnest with the English chemist John Dalton. In 1803 he proposed that the known facts of chemistry could be explained if one assumed that all matter is ultimately comprised of units known as atoms which were then thought to be the smallest particles in the world. No one was sure at the time whether atoms really existed or not, but they were admitted as being highly convenient entities that helped to systematise the science of chemistry. Thus atoms of hydrogen, nitrogen, oxygen, carbon and so on came to be regarded as fundamental units of which the physical universe is comprised. The science of chemistry was left to unravel which particular combination of atoms occurred in a given material, as in water where two hydrogen atoms are linked to an oxygen atom to form a water molecule. And it is the precise combinations of these atoms within molecules that determine the characteristic chemical behaviour of different substances in the world.

Early chemists regarded atoms as being tiny spheres, too small to be seen in an optical microscope, but no one was quite sure how small they were. Today we know that some 100 million atoms placed side by side would make up a distance of about a centimetre.

No chemist, even in the latter decades of the nineteenth century, thought it would ever be possible to probe into the atom and determine its content. Then, one fine day in 1896, the foundations of atomic physics came accidentally to be laid. The French physicist Becquerel was working on the properties of the recently discovered X-rays when quite by accident he stumbled on the phenomenon of radioactivity. He discovered that material containing uranium atoms gave off a type of radiation that could fog photographic plates, but these rays he showed could not be X-rays. They were essentially uranium atoms ejecting so-called alpha particles, which were later found to be identical to the helium nucleus. Becquerel had discovered radioactivity, as it came to be called. From then on it did not take long for scientists to discover that atoms themselves were complex assemblies of smaller particles, negatively charged electrons and a compact nucleus containing positively charged protons. Electrons and protons thus had charges equal in magnitude but opposite in sign. Years later in 1932 when the neutron was discovered it was found to be electrically neutral.

Following the discovery of radioactivity it soon became clear that the nuclei of different atoms had different electric charges because they were made up of different numbers of protons. In atoms that were electrically neutral the number of protons in the nucleus had to balance the number of electrons around it. Protons and neutrons have approximately, though not exactly, the same mass, while the electron is 1836 times lighter. A somewhat crude model of an atom is to think of it as a miniature solar system, with a nucleus built from neutrons and protons representing the central 'Sun', and with electrons representing identical planets in orbits around the 'Sun'. A crucial difference, however, is that atoms are proportionately larger with respect to the size of the nucleus than the planetary system is with respect to the size of the Sun.

In terms of the three building blocks of matter, electrons, protons and neutrons, the atomic nucleus turned out to be a composite structure for all elements except hydrogen. For hydrogen, the most abundant element in the Universe, the nucleus is simply a single proton. Nuclei of heavier atoms contain several protons and neutrons practically jostling against each other, posing the urgent question as to how particles of like charge (protons) hold themselves together and are not flung apart as they would be under the effect of electrical forces alone.

There seem to be four principal types of force that shape the Universe: gravity, which operates on an astronomical scale; electromagnetism, which holds negatively charged electrons to positively charged nuclei; the weak nuclear force, which controls the radioactivity of certain nuclei; and the strong nuclear force, which holds protons and neutrons together within the atomic nucleus. The so-called coupling constant for the strong nuclear force is about a thousand times greater than that for the electromagnetic force.

In 1935 the Japanese physicist Hideki Yukawa suggested that if such a strong nuclear force existed, it might be maintained essentially by a steady exchange of particles of intermediate size (mesons) between the jostling neutrons and protons within the nucleus. Earlier in 1931 the Austrian physicist Wolfgang Pauli had suggested that in order to explain certain facts about electrons, another kind of particle, the neutrino, possibly with zero mass, must also exist. So to electrons, protons and neutrons, neutrinos and mesons were added as theoretical concepts to be looked for. Also there were predictions of antiparticles – antineutrino, positron and so on – extending considerably the complexity of the subatomic world.

Attempts to observe mesons required the capacity to bash atomic nuclei with charged particles of enormously high energies. To produce these energies enormous particle accelerators comprised of rings of giant electromagnets came to be constructed at great cost. Yet, in the early years, terrestrial man-made accelerators were hard put to compete with naturally occurring accelerators in our galaxy or in external galaxies. These naturally occurring accelerators could yield atomic particles with energies that were far higher, measured in thousands of billions of electron volts. Such particles, the so-called cosmic rays, are comprised of bare atomic nuclei that rain down on the terrestrial atmosphere at a steady rate. The impinging primary particle collides with air molecules to produce a shower of particles which sometimes include esoteric fundamental particles that are generated by the high energies involved.

The pi-meson was discovered in this way by the English physicist C.F.Powell in 1947. In the same year, G.Rochester and C.Butler found evidence for the existence of a heavier meson, the so-called K-meson. Since then an impressive list has developed of other basic particles which have been discovered both in cosmic ray studies and from experiments using fast accelerators. Many of the rarer types are difficult to detect for the reason that they have very short lifetimes, in some cases less than a billionth of a second, after which they decay into another form of particle or particles.

Now it seems that all these fundamental particles are themselves comprised of more basic particles known as 'quarks'. There appear to be at least five types of quarks that are needed in order to account for all the particles that have actually been seen. Protons and neutrons are made of two types of quark called 'up' and 'down'. The proton is two 'ups' and one 'down'; the neutron is two 'downs' and one 'up'. Mesons consist of a quark paired with an antiquark. In order to describe a quark completely one needs to specify spin, flavour and colour from a total minimally of 30 different combinations. It is hoped that by combining these possibilities in various ways all the fundamental particles in the Universe can be generated.

Developments in probing the nature of fundamental particles have been matched by our efforts to explore the wider Universe. Until relatively recently the only astronomical observations available were those made through the narrow optical window (3000–8000 Å) which represented only a small fraction of the total electromagnetic spectrum. Over the years astronomers had learnt to improve their techniques for collecting this optical radiation and to scrutinise it with ever-increasing thoroughness. From optical studies we had learnt, for instance, that the Sun is a single star in a Galaxy comprising some 100,000 million stars, and that about 1000 million each comprised of similar star systems populate the observable Universe.

Besides the visual spectrum, other regions of the electromagnetic spectrum began to be opened for astronomical investigation from 1945. The science of radar, developed during the Second World War, was turned to the benefit of astronomy, leading to the science of radioastronomy. Further dramatic discoveries followed the opening up of other windows in the electromagnetic spectrum, which occurred in the early 1960s and has continued into the present decade. Developments in solid-state physics and cryogenic techniques enabled the construction of infrared detectors which span the wavelength range 1 to 100 micrometres. Such equipment used in conjuction with telescopes carried on satellites (e.g. IRAS) and at well-chosen ground-based observatories (e.g. the Anglo-Australian Observatory) are continuing to provide a wealth of new information. Much of the infrared data obtained in this way has had a profound relevance to the thesis of this book, as we have seen in earlier chapters. Ultraviolet observations of stellar spectra also became possible with the dawn of the space age in the 1960s. Then followed the advent of X-ray and gamma ray astronomy. With several satellites dedicated to these various goals a flood of new discoveries has followed.

One of the most important milestones of observational astronomy that lies in the very near future is the launching and operation of the 2.4 metre (94 inch) Large Space Telescope, which is currently scheduled for launch in December 1988. For the first time the fullest potential of a large, high resolution optical telescope will be used free of the obscuring and distorting effects of the terrestrial atmosphere. A high degree of pointing accuracy as well as sensitivity over a wide wavelength range is expected to be achieved. The telescope which will orbit around the Earth will surely probe further and deeper into space than any existing ground-based telescope could do. Amongst the important astronomical questions that could eventually be settled is the matter of planets around nearby stars. The Space Telescope is believed to have a sensitivity capable of detecting such planets.

Today, astronomical objects are studied over some 13 decades in wavelength, ranging from gamma rays to long-wave radio pho-

tons. Objects in the Universe radiate and absorb over this spectral region. The radiation could be of thermal origin – emission
from solid bodies and particles, neutral and ionised gases heated
to various temperatures; or it could be non-thermal in character,
for instance arising from high energy electrons spiralling in magnetic fields. Radiation could occur over a continuum of wavelengths or at discrete wavelengths in the form of spectral lines
over much of this region.

The Sun, being the nearest star to us, has naturally been the
subject of close scrutiny over many years. The 11-year sunspot
cycle, the periodic reversal of the Sun's magnetic field and the
detailed behaviour of the solar wind, still present challenging
problems to astrophysicists. More significant for the story of
cosmic life is the evolutionary history of the Sun seen as a star
changing with time. The emergence and maintenance of life on
planetary objects are matters that are intimately connected with
that history.

To obtain information on how the Sun and other stars evolve
– i.e. change their surface temperature, radius, intrinsic brightness, and composition – astronomers have made careful studies
of groups of stars known as clusters. Each cluster represents
a collection of stars born essentially at the same time, but containing a spectrum of star masses, presenting therefore a snap-
shot image of a phase of evolution of a group of stars.
Investigations of the properties of stars in such clusters provide
the basic raw data relating to the evolution of stars. Addition-
ally, however, our present understanding of stellar evolution
would not have been possible but for the developments in
nuclear physics that we referred to earlier in this chapter.

We shall now sketch briefly the evolution of a star with a
mass similar to that of the Sun. The star begins life as a fragment
of an interstellar cloud containing by mass about 74 per cent
hydrogen, 24 per cent helium, about one-and-a-half per cent carbon, nitrogen and oxygen, and the remaining half per cent or
so consisting of heavier elements like magnesium, silicon and
iron. The hydrogen and helium exists as gas, but the rest exists
mostly as dust particles, bacteria according to our point of view.
The interstellar cloud fragment contracts under its own self-
gravitation, heating up as it does so. The temperature at the

centre of such a 'protostar' continues to rise, and the contraction is not halted until a temperature of about 10 million degrees is reached. The intermediate stages of contraction are thought to be swift, but the last stages last for some 20 million years, during which extended period planets and comets are formed. The first set of thermonuclear reactions, resulting in the fusion of 4 hydrogen nuclei into a helium nucleus, then begins.

As is well known, the energy released in these reactions provides the main power source in the Sun and most other stars throughout the major part of their lives. For the Sun this highly stable phase is estimated to last about 10,000 million years; and we, at the moment, are well settled in the middle of this phase. The end of this phase would be heralded when a core of helium amounting to about 10% of the mass of the star is formed at the star's centre. The star then rapidly distends its outer layers, and the surface temperature drops below 3000 degrees Kelvin when the radius has become enlarged by a factor of about 100. The star becomes a red giant. Hydrogen burning continues for a while in a shell around the central helium core. This is later accompanied by further contraction and heating of the core and by the onset of nuclear reactions involving the conversion of helium to carbon, oxygen and heavier elements. In the giant phase convective motions are set up in the outer layers of the star, and, on account of the low surface gravity that prevails, material may be able to escape from the star. There is observational evidence of gas flows occurring from the surfaces of such red giant stars. Mass flows of this kind occurring in the first generation of stars to form within a galaxy would have provided the carbon, nitrogen, oxygen and heavier elements which serve as the feedstock of cosmic microbiology.

The subsequent evolutionary story of a star is still more complicated and rather strongly dependent on the precise value of the initial stellar mass. The general course of events can be described by a succession of nuclear reactions – hydrogen burning, helium burning, and so on – taking over as the main power source of the star. Eventually for a star like the Sun, partly through mass loss in the giant phase and partly through the nuclear reactions themselves, the power source becomes exhausted and the star settles at last into the highly compact form known as a white dwarf. But for stars of large mass a cataclysmic event known as a supernova explosion intervenes, in which event more biological feedstock becomes generated and dispersed. These ideas, which are now widely accepted, followed from the pioneering work of one of us (FH) in collaboration with W.A.Fowler and Geoffrey and Margaret Burbidge in the middle years of the present century.

Compared with the precise nature of the conclusions that could be drawn from studies in stellar physics, inquiries into the nature of the Universe at large – the study of cosmology – have led to indefinite results and to the emergence of religious-type beliefs. The first piece of hard evidence that had a bearing on the structure of the Universe as a whole came in the first third of this century through the work of American astronomer Edwin Hubble. The picture of an expanding universe emerged, a Universe in which all galaxies seemed to be moving away from each other at speeds that increased with their distance apart.

The observations of Hubble, that distant galaxies were rushing away from us, might for a brief moment have suggested a return to a pre-Copernican Universe, with our galaxy being placed at the centre of things. This potentially uncomfortable situation did not in fact arise, for a resolution of the apparent difficulty soon emerged from Einstein's theory of relativity, which of course had to be considered when dealing with problems that encompassed the whole Universe. The upshot of these considerations was to construct a model of the Universe in which every single galaxy was rushing away with respect to every other. A simple analogy that is sometimes given is of equally spaced dots on the surface of a spherical rubber balloon, a balloon that is being blown up from inside. The dots behave like Hubble's galaxies. From the perspective of any particular dot (galaxy) the other dots (galaxies) expand away from it. Of course the world of galaxies is not two-dimensional like the dots on the balloon, but cosmologists can argue by precise mathematics that if our real world expanded analogously to the balloon, it would lead to a Universe similar in character to that which is observed.

If we imagine the process to be reversed, turning the expansion into contraction, it might appear that the Universe was crowded together into a 'singularity' some 15 billion years ago, suggesting an origin at a definite moment in time, thus giving rise to the 'Big Bang' model of the Universe which was first suggested in 1922 by the Russian mathematician Alexander Friedman.

In 1948, however, one of us (FH) together with Thomas Gold and Hermann Bondi argued that the available astronomical evidence did not necessarily imply the correctness of this Big Bang model. It was suggested that the Universe might always have existed in much the same form as it is at present. There was no logical requirement for either a beginning or an end, thus giving rise to a picture of the Universe that was essentially in a 'Steady State'. A consequence of this model was of course that matter had to be created from a field of energy in such a way as to fill the recesses that are caused by the expansion of the Universe. As the galaxies expanded and moved further apart from one another, 'new' material would be required to fill the vacant spaces if an overall appearance of a Steady State Universe were to be maintained. The typical creation rate required was estimated as the production of the equivalent mass to one hydrogen atom in a typical assembly hall every 1000 years, only a very slow average rate.

From the 1950s onwards these two cosmological models – the Big Bang model and the Steady State model – came under increasingly close scrutiny. In the early 1960s studies of the most distant galaxies using radiotelescopes were claimed to indicate that the Universe was slightly more compacted in the past than it is at the present time, thereby favouring, albeit marginally, the Big Bang model. More recent observations, however, have indicated that this does not seem to be so.

An important observation that has not, in itself, proved wrong was the discovery by Arno Penzias and Robert Wilson in 1965 of a background of low-energy microwave radiation filling the Universe. The temperature of this background was determined as about 2.7 degrees Kelvin, and the radiation appeared to come with almost the same intensity in all directions. This was what has come to be called the Cosmic Microwave Background. The interpretation that was immediately put on this new datum was that it was a relic of the heat left over from the explosive origin of the Universe some 15 billion years ago. It had to be so in the minds of most cosmologists, who by now have become convinced that the Big Bang was almost as much fact as was the proposition that the Sun was the centre of our solar system. Thus cosmologists came to regard the Cosmic Microwave Background as the final conclusive proof of the Big Bang. Yet, as we shall see, this was far from conclusive proof. The existence of the background itself is indisputable as a fact relating to the Universe, but its interpretation as relic radiation from the Big Bang is neither proved, nor is it free of intrinsic faults. In particular, the background is amazingly smooth in its distribution across the sky. If it had been generated in an initial Big Bang, the later formation of clusters of galaxies should have left an observable irregular signature on it. Such a signature has been looked for with great care but has not been found. The implication is that the background has been smoothed and perhaps generated, not at an origin of the Universe, but subsequent to the condensation of galaxies.

From 1968 onwards, we ourselves, working in collaboration with Vincent Reddish and Jayant Narlikar, have pointed out that the microwave background might be explained if starlight in galaxies is absorbed and re-emitted as microwaves by dust particles that possess suitable radiative properties. We have searched for some time for suitable types of particles, and recently we have shown that dust comprised of long slender needles or whiskers made largely of the element iron might fit the bill. Such particles form quite readily in the condensation of metallic vapours under laboratory conditions. When supernovae explode it is known that explosive synthesis of nuclei leads to the production of iron gas, and we can show using available laboratory data that a substantial fraction of this iron will be expelled into space in the form of slender whiskers of diameters 0.01 micrometres and lengths exceeding 1 millimetre. The absorption by such grains of radiation at microwave frequencies could be a thousand times more efficient than in the optical and infrared. Consequently, if an appreciable fraction of iron produced in supernovae exists as whiskers in intergalactic space the Universe could be opaque over cosmological distance scales at radio and microwave frequencies but remain translucent in the visible region of the spectrum. Metallic whiskers would thus be highly effective in thermalising the optical radiation from galaxies to yield a smooth microwave background, as is observed, provided it could first be degraded into infrared radiation. There is of course ample evidence to show that the latter is no problem, so at least one erstwhile prop of Big-bang cosmology is seen to be insecure.

As far as models of the entire Universe are concerned, the datum that is usually construed as evidence in favour of the Big Bang theory is thus accountable in another way. Thus we find that the most popular model of the Universe, which is at the present time being maintained with an almost religious fervour, is far from being proved. Nor is any competing model proven, at least for the time being.

NINE

The control of galaxies

We have seen in earlier chapters that our attempts to probe the physical Universe have led to many aspects of it eventually emerging with great clarity. For instance, our analyses of the properties of cosmic dust have ultimately defined a particle model that cannot be allowed to depart from the known properties of bacteria to any sensible degree. Similarly, the application of nuclear physics to astronomy has led, as we saw in the previous chapter, to precise calculations of nucleosynthetic processes in stellar interiors; and these in turn have given us an accurate knowledge of how stars evolve. In sharp contrast, there are other areas of astronomy where the understanding of even the basic processes involved have remained elusive, resulting in a fuzzy state of knowledge, despite considerable effort and talent that have been directed to such studies. Almost invariably these apparently enigmatic astronomical phenomena seem to be wedded in some way to the properties of cosmic dust grains – bacteria in our view. The recognition of an appropriate role for bacteria, or of living systems derived from bacteria, seems to be a possible way forward for ultimately resolving these apparently intractable astronomical problems. We shall argue here that cosmic bacteria, or superstructures built from them, may well be in control of our galaxy as a whole, and even of external galactic systems.

The association of extremely young stars with clouds of gas and dust has been recognised for several decades. An exciting outcome of infrared astronomy has been the discovery of strong infrared radiation, particularly in the waveband 2–20 micrometres from shells of dust grains that surround many stars. A significant proportion of these sources have spectra that peak near the wavelength 10 micrometres, indicating that the bulk of the associated dust is radiating at a temperature of about 300 degrees Kelvin. This temperature is of course ideal for the maintenance of living systems, and it is entirely possible that these protostellar systems are intelligently contrived so as to be radiating at this optimal wavelength for life.

Extremely young supergiant stars and a group of stars known as the T-Tauri stars exhibit erratic variations in light intensity that are not yet fully explained. It is believed that in many such cases we may be seeing cocoons of dust that are left behind after stars have been formed. It seems to us entirely likely that, in some instances at least, these objects resemble what are known in the trade as 'Dyson Spheres', objects that conceal perhaps an even higher intelligence than our own. The suggestion is that an intelligence of some kind has become assembled and ordered from the basic bacterial and viral building blocks to reach high enough levels of sophistication for harnessing the energy of the embryonic star. The cocoons of dust that now shroud these embryonic stars could be cleverly designed to trap and utilise essentially all the optical radiation from a parent star, and so maintain a sizable enclosure in which life could develop. Solid objects of planetary size within such an enclosure would then be maintained at a temperature optimal for life.

It is relevant to note in this connection that the phenomenon of masering (the microwave analogue of lasering) is known to operate in many molecular clouds that are associated with young stars. The hydroxyl radical and the water molecule, which are principally involved in producing this phenomenon, are perhaps quite significantly molecules that are obviously crucial for life. In a maser the upper level of a microwave transition becomes systematically over-populated, leading to the emission of coherent radiation at a sharply defined frequency. The general belief among astronomers is that this effect is caused by some purely physical process that operates in the tenuous conditions prevalent in a molecular cloud, where population inversions are caused by collisions with hydrogen molecules or through the absorption of radiation. These astronomical masers were for a brief while thought to be intelligent devices, but nowadays a purely physical explanation is considered to be preferable. Needless to say, there are several problems concealed within these so-called physical models, and the involvement of a living system is by no means ruled out. On the contrary, the involvement of a living system might be highly probable. After all, it is in the very nature of living systems to be able to produce 'population inversions' in molecules because of the highly 'non-equilibrium' conditions that prevail within cells. We suspect that the story in this regard may well be a good deal more complex and intriguing than it has been customary to admit.

The pre-existence of dust grains is widely believed to be an absolute requirement for the formation of normal stars in the mass range of 2/10 to 50 times the mass of the Sun from an interstellar cloud that is many times more massive still. The dust particles are believed to play a role in determining the processes by which such a larger cloud breaks up into cloudlets from which stars can condense.

According to conventional astronomy, the original far larger cosmic cloud that contracted to form our entire galaxy would not have contained any more than a minute fraction of the dust required to produce extensive fragmentation. Nor would it have contained any significant amount of heavy elements, such as carbon, nitrogen and oxygen, from which dust particles may have readily formed. Such elements are produced by nuclear processes in stars, in the absence of which they would certainly have been in short supply. It then seems inevitable according to this point of view that the collapse of a pre-galactic cloud would lead to the condensation of much larger units that were unable to break up immediately into smaller pieces. The result might well be the formation of supermassive stars, each of which could have masses exceeding that of the Sun by a factor of 100,000. Stars of this sort were first proposed by one of us (FH) and W.A.Fowler in 1964. The evolutionary history of a supermassive star is expected to be a speeded up version of the evolution of a normal star, ending in an explosive injection of heavy elements and inorganic dust grains into space.

Explosions of this kind would provide the material necessary for the formation of the first generation of ordinary stars, and also the chemical feedstock needed for the rapid amplification of a small initial component of cosmic biology with which our galaxy may conceivably have been endowed. Some of this initial endowment that remained in the outer periphery of a typical galactic cloud would remain in a cold viable state until the chemical feedstock was eventually provided by nuclear processes in supermassive stars. The subsequent evolution of life in our galaxy would be controlled by processes that occur in later generations of stars and by the cosmic life cycle shown in an earlier chapter (Fig. 4.5).

The existence of dust in external galaxies has been known for many years. Fig. 9.1 shows the Andromeda nebula, a spiral galaxy similar to our own, nearly 2 million light years away. This is a collection of about 100 billion stars arranged in spiral arms intermixed with conspicuous dust lanes. The entire system is thought to resemble our own galaxy to a remarkable degree. If dust grains are to be equated with biology, as we seem forced to suppose in our own galaxy, it would seem logical to conclude that the Andromeda nebula must also be teeming with life. The dust grains in external galaxies, wherever they could be probed, for example in the Magellanic Clouds, show properties in the visual spectral region that are almost identical to interstellar dust in our own galaxy. The inference that dust in external galaxies must also be biological is therefore based on as firm a foundation as there could be, in view of the present state of our knowledge. Fig. 9.2 shows an even more conspicuous dust lane in the spiral galaxy known as the Sombrero Hat, viewed edge on at a distance of nearly 40 million light years. There are numerous examples of a similar kind, and also some even more dramatic cases of entire galaxies being enveloped in gigantic halos of dust.

Dust particles sometimes show up in extragalactic sources by virtue of the infrared radiation they emit. A class of infrared-emitting spiral galaxy that has recently attracted much attention is that of the so-called Seyfert galaxies. A property common to most spiral galaxies is that the spectra of their integrated light (the light of all the stars lumped together) resemble spectra of the stars in our own galaxy. In 1943 Carl Seyfert noted that

Fig. 9.1 The Andromeda nebula. (Courtesy of Mt Wilson and Palomar Observatories)

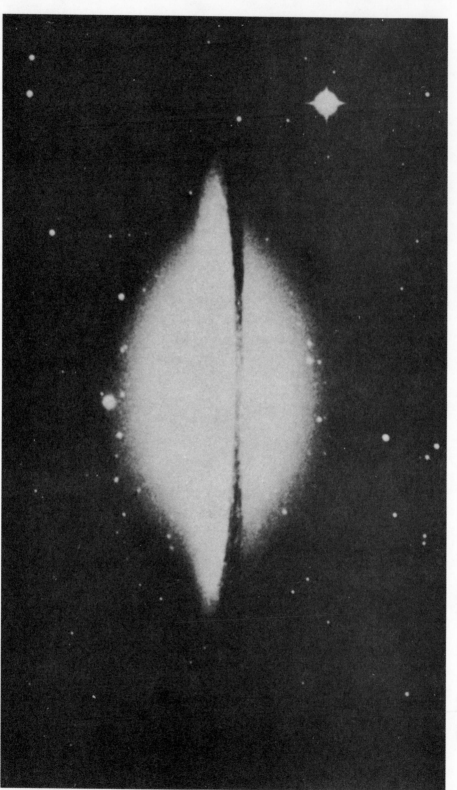

Fig. 9.2 Sombrero Hat galaxy NGC4594 in the constellation of Virgo. This is a spiral galaxy viewed edge-on showing a conspicuous dust lane through the central plane. (Courtesy of Mt Wilson and Palomar Observatories.)

about 1 per cent of all spiral galaxies departed from this trend towards their centres. Much of the light emitted from this peculiar minority group comes from a compact central region and their spectra are markedly different from normal stars, indicating high levels of excitation and agitation of gaseous atoms. In 1968 F.J.Low and his collaborators discovered that many Seyfert galaxies are strong emitters of infrared radiation in the wavelength range 2–20 micrometres. The observational data seemed to indicate that this infrared radiation could only arise from a shell of dust grains heated by an intense central source of ultraviolet light. The dust grains emitting at the 2 micrometre wavelength had to be so hot (over a thousand degrees) that the only plausible composition was carbon in the form of graphite. This dust could not be in the form of viable biological particles for the reason that the temperatures are too high. It appears most likely that these central sources are indeed collections of supermassive stars of the kind we have already discussed, or of the collapsed residues of such stars. Explosions of such objects could have provided both the inorganic carbon dust as well as the energetic photons needed to heat the dust. The galaxy M82 is a case where an explosion at the centre has led to the injection of vast quantities of dust particles into intergalactic space (Fig. 9.3).

The likely interpretation of the Seyfert phenomenon, and indeed of the more general phenomenon of 'active galaxies', is that they are similar to the earliest phases in the evolution of a spiral galaxy. It is in the early phases that the chemical feedstock needed for a large-scale explosion of biology in the galaxy will be generated. Whether such phenomena may themselves be ultimately controlled by biological processes is an interesting question that could be asked. If, as we believe, incipient biology must already exist in some form, it would make sense for such biology to contrive to use the enormous energy output of the Seyfert galaxies (much greater than the luminosity of a normal galaxy) to serve its own ends. Could bacterial cells have become elaborated even at this early stage into structures possessing high levels of intelligence, occupying perhaps a small number of planetary or cometary objects at a suitably congenial distance away from the central nucleus?

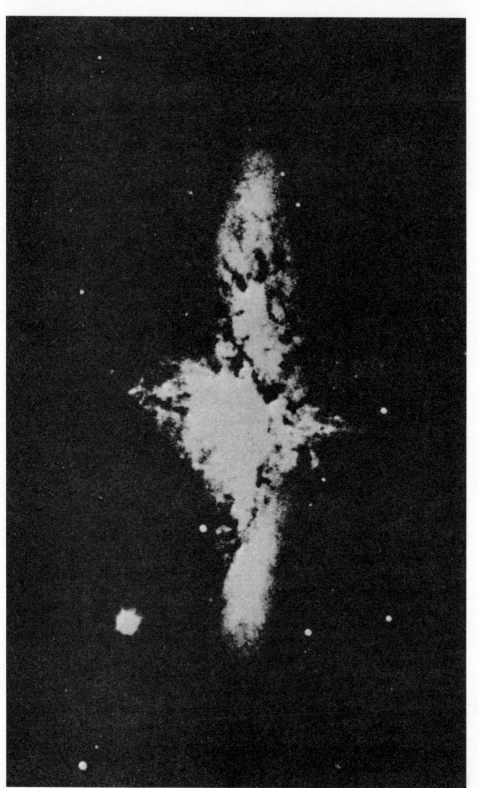

Fig. 9.3 Exploding galaxy M82. (Courtesy of Mt Wilson and Palomar Observatories)

Returning to our own galaxy, and to the bacterial particles within it, we cannot afford to omit a reference to their connection with the galactic magnetic field. A magnetic field equal in strength to about one millionth part of the magnetic field of the Earth exists throughout much of our galaxy. This magnetic field can be used to trace the spiral arms of the galaxy, following a pattern of field lines that essentially connects associations of young stars and dust (the so-called 'OB associations') in which new stars are forming at a rapid rate. A curious property of the interstellar dust grains is that they not only dim starlight, but also polarise it. There is a partial linear polarisation of the light of any star that is dimmed by the dust. The implication is that at least a fraction of the dust particles have elongated shapes (similar to bacilli) and they are systematically lined up in our galaxy. Moreover, the direction of the alignment is such that the long axes of the dust tend to be at right angles to the direction of the galactic magnetic field at every point. This alignment is not a rigid or static alignment as of iron filings in the field of a bar magnet. The grains in space are in a state of rapid spin due to collisions with surrounding gas atoms, so a stochastic or average tendency for lining up to a slight degree is required. A friction-like interaction of the spinning dust with the magnetic field appears necessary to produce the desired effect. Such magnetic interaction is not possible unless the particles possess appropriate magnetic properties. However, all the well-known magnetic properties of inorganic candidate materials for the dust grains have proved, over the years, to be inadequate for this purpose. Yet a class of magnetic bacteria which exist on the Earth is known to have exactly the right kind of property (superferromagnetic domains) to effect the degree of alignment that is observed. This might sound like a small detail compared with the much wider issues we have discussed so far, but to us it served as yet another powerful confirmation of the life-from-space theory.

It is interesting that issues which would inevitably influence the well-being of life on a cosmic scale involve problems in astronomy that have seemed intractable. Included amongst these is the problem of the formation of new stars and its relation to the cosmic dust grains. The major unresolved issues are as follows:

What decides the rate at which stars form from the interstellar gas?

When they are formed, what decides the mass distribution of the stars?

What decides the rotations of the stars, and whether they are formed with planetary systems?

What decides how star formation is correlated throughout a whole galaxy, leading to the production of both grains and stars on a far-flung basis, often with the appearance of a new set of spiral arms for a galaxy?

The answers to these questions are almost certainly connected with the existence of a magnetic field everywhere throughout our galaxy, a magnetic field that we saw has an intimate connection with the bacterial dust grains. But the nature and origin of the galactic magnetic field is a further unresolved problem, and so the additional question must be added:

How did the magnetic field of our galaxy come into being?

This further question has proved so baffling that many astronomers have given up hope of answering it, by claiming the magnetic field to be truly primordial, it being imposed on the Universe at the moment of its origin or generated in its very early history. The magnetic field is what it is, because it was what it was, right back to the first page of Genesis. If this view is correct the situation is highly unsatisfactory, because it makes so much of what we observe at present contingent on situations that cannot be observed and which therefore at best are speculative.

To understand the difficulty of the problem, suppose we were to try to generate the magnetic field of our galaxy with the aid of a fictitious electric battery having one terminal connected to the centre and the other to the outside of our galaxy. With the battery switched on, an electric current would begin to flow through the interstellar gas. Owing to a phenomenon known as 'inductance', the electric current would at first be exceedingly small, but as time went on the current would become stronger, and as it did so the magnetic field associated with the current would increase in its intensity. Suppose we allow the current to grow for the whole age of our galaxy, about 10,000 million years. What voltage do we need for the battery so that after

10,000 million years the resulting magnetic field will be as strong as the galactic magnetic field that is actually observed? The answer is about 10,000,000 million volts. What process, we may ask, could have produced a battery of such enormous voltage that could operate for a time as long as 10,000 million years?

A conceivable answer is a stream of electrically charged interstellar dust grains projected at high speed, 100 km per second or more, into an electrically neutral gas, a process in some respects similar to that which drives a terrestrial thunderstorm. The required projection speeds of 100 km per second are attainable by radiation pressure, the pressure generated by light waves striking the surface of a dust grain. The remaining problems are first to attain systematic directivity for a stream of grains, and second to maintain electrical neutrality in the gas through which the stream passes (when neutrality fails in a terrestrial thunderstorm there is a lightning flash and the electric battery is instantly discharged). With non-biological grains it is difficult, if not indeed impossible, to resolve these issues. Bacteria, on the other hand, have far more complex properties than inorganic dust grains, and may be able to exert a covert control both on stream directions and on the neutrality of the interstellar gas. The issue is not proved, but it is conceivable, and if it were to happen, bacteria would be well placed to control the whole process of star formation. The fact that magnetic bacilli actually appear to be connected to the magnetic field lines that thread through the spiral arms, connecting one region of star formation with another, goes a long way towards establishing the plausibility of such a point of view.

The nutrient supply for a population of interstellar bacteria comes from mass flows out of the large galactic population of some 100,000 million old stars. 'Giants' arising in the evolution of such stars experience a phenomenon in which material containing nitrogen, carbon monoxide, water, hydrogen, helium, some refractory solid particles and supplies of trace elements essential for life flows continuously outward into space. In total from all giant stars, a mass about equal to the Sun is expelled each year to join the interstellar gas. This is the nutrient supply for cosmic biology.

The problem for interstellar bacteria is that the nutrient supply cannot be converted immediately into an increase of the bacteria population, because of the need for liquid water, which cannot exist at the low pressures of interstellar space. Water in interstellar space exists either as a vapour or as solid ice, depending on its temperature. Only through star formation, leading to associated planets and smaller cometary bodies, can there be access to liquid water. To control the conditions leading to star formation would therefore be of paramount importance to cosmic biology.

Conditions suited to the presence of liquid water can exist for long periods of time on planets like the Earth. Liquid water need not exist, however, for long periods of time, since bacteria can multiply so extremely rapidly given suitable conditions. Shorter periods could exist on bodies much smaller than planets, and in the early high-luminosity phase of newly formed stars these water-carrying bodies could lie far out from the stars. In the case of our own solar system, liquid water could quite well have persisted in the early days far out towards the periphery, and it could have existed at the surfaces of bodies of lunar size or inside still smaller bodies like cometary nuclei. The nutrients present in the outer regions of the solar system must have exceeded by many billions of times the amount at the Earth's surface. Hence the short-lived conditions associated with star formation must be of far greater importance to the population of interstellar bacteria than the long-lived, more or less permanent environment offered by planets like the Earth.

It has long been clear that the detailed properties of our own solar system are not at all what would be expected for a blob of interstellar gas condensing in a more or less random way. Only by a rigorous control of the rotation of various parts of the system could such an arrangement like ours have come to pass. The key to maintaining control over the rotation seems to lie once again in a magnetic field, as indeed does the whole phenomenon of star formation. The surest way for interstellar bacteria to prosper in their numbers would be through maintaining a firm grip on all aspects of the interstellar magnetic field. By doing so they could control in a sensitive way not only the rate at which stars form but also the kinds of star system that are produced.

For many years astronomers have been used to thinking of star-forming episodes as being just incidentally, and somewhat pointlessly, accompanied by the production of dust clouds. From astronomical observations we know that these episodes are sometimes local but they are often galaxy-wide. They are thought to be triggered by some large-scale event (a galactic 'spiral wave') the after-effects of which persist for some considerable time, several hundred million years. The formation of the exceptionally bright stars which delineate the spiral structures of galaxies has often been associated with these episodes. From the arguments in this chapter it seems that even the origin of the conspicuous spiral structures of galaxies may well be biological in its nature.

TEN

The concept of a Creator

The position we have now reached is that the components of terrestrial life are dispersed in the form of cosmic dust particles in the most far-flung places in the Universe. As a simple analogy we can regard the genetic molecular arrangements that correspond to crucial characteristics of living things as pieces of an open-ended jigsaw puzzle that could, when pieced together, give rise to an astronomically wide range of possible life-forms. Such pieces, from which all the known terrestrial life-forms might be imagined to be derived, including those coding for intelligence, are then scattered everywhere as part of the fabric of galaxies. The individual jigsaw pieces could be interpreted as informational units – e.g. the information needed to code for eyes, feathers, intelligence, and so on. All that is left to chance at any particular assembly site like the Earth is for these units to be allowed to come together in appropriate habitats, where feasible genetic combinations would be continually sorted out through the process of competition for survival in local environments.

It has often been asked by our critics whether, in arriving at this picture, we have really made any progress towards understanding the ultimate origin of life. Have we not, it is asked, merely pushed back this problem to a more remote place in the Universe, and to a more distant time in the past? In some ways this comment could be seen as valid. But it should be stressed that even such a shift of venue for creation was not a matter of fanciful or arbitrary choice, but it was dictated by hard facts. If life did not emerge on the Earth in the traditional primordial soup then it becomes a matter of great scientific importance to recognise this fact. We have argued in earlier chapters that such must be the case beyond reasonable doubt. Life on Earth must have come here in the form of genetic packages that came to be progressively assembled into the multitudinous array of shapes and forms that we find today. Biological effects such as mutations, gene doublings, re-combinations, and so on, to which the whole of the evolutionary process is normally attributed, can in our view serve no more than a fine tuning to be superposed on the much greater cosmic assembly process.

The outstanding question that remains to be answered is the origin of the information content within the cosmic genetic packages that contributes to the evolution of life. This is what philosophers would regard as the quest for a First Cause. Did such information arise from a purely random assembly process, somewhere in the Universe, or was the cosmic genetic system in some way deliberately created? To attempt an answer to this wider question let us turn to a group of molecules known as enzymes. Enzymes are polymers or chains of smaller units known as the amino acids. There are enzymes to assist almost every basic biochemical process and without such enzymes biology as we recognise it cannot exist. There are, for instance, enzymes to assist digestion, there are enzymes that zip apart the larger molecules of life into their component parts, and there are enzymes that assist in the assembly of the smaller molecules into long chain polymers, such as proteins and nucleic acids. In all, 2000 or more enzymes are crucial across a wide spectrum of life, ranging from simple microorganisms all the way up to Man. In general there is so much correspondence in the arrangement of amino acids within an individual enzyme, across the whole spectrum of life, that it is sometimes possible to use the enzymes of microorganisms to serve Man.

Every individual enzyme among the 2000 or more depends for its action on the way that the 20 biological amino acids are arranged along the polymer chain in question. A considerable fraction of positions in the chain have of necessity to possess a particular member of this set of 20 amino acids. Minimally, one could say that correct choices of amino acids must occur at 15 sites, while for many enzymes the number of obligatory choices is much larger than this. A simple calculation then shows that the chance of obtaining the necessary total of 2000 enzymes by randomly assembling amino acid chains is exceedingly minute. The random chance is not a million to one against, or a billion to one or even a trillion to one against, but p to 1 against, with p minimally an enormous superastronomical number equal to $10^{40,000}$ (1 followed by 40,000 zeros). The odds we have thus computed are only for the enzymes, and of course correct arrangements within many other important macromolecules of life besides enzymes must also be considered. The molecules histone-4 and cytochrome-c are two such examples, each with exceedingly small probability of being obtained by chance. If all these other relevant molecules for life are also taken account of in our calculation, the situation for conventional biology becomes doubly worse. The odds of one in $10^{40,000}$ against are horrendous enough, but that would have to be increased to a major degree. Such a number exceeds the total number of fundamental particles throughout the observed Universe by very, very many orders of magnitude. So great are the odds against life being produced in a purely mechanistic way that the difficulties for an Earthbound, mechanistic biology are in our view intrinsically insuperable.

Such criticisms of this conclusion as have been voiced are, in the main, of a superficial polemical kind. By prevailing cultural standards it is usually thought that this type of conclusion is so outrageous and unacceptable that it has to be fought and condemned at all costs. Facts themselves cease to be important, and it is considered permissible to pile unlikely hypotheses, one upon another. If there remains even the slightest chance of maintaining the *status quo* within some type of quasi-logical framework the situation is that anything goes. The hypotheses that come to be invoked take on an extremely complex character, which appears to remain invisible to the protagonists of conventional theory.

It has been said that the enzymes themselves have 'evolved' from unknown precursor systems that were much simpler, and perhaps much more likely to form through random processes.

We have no knowledge that any such precursor systems ever existed, and if they did the crucial question still arises of whether such a system could have defined a random evolutionary trend that led eventually to the 'discovery' of the present enzyme system. If one forces the logic to assert that life must have evolved in a purely mechanistic way, then one is led to suppose that matter has an inherent tendency to find a final highly reproducible order within the crucial life molecules. In other words, the orderings necessary for life are required to be built into the properties of matter, perhaps at a subatomic level. But the circumstance that not even the slightest hint of such a tendency has been discovered in laboratory experiments on the origins of life casts doubt on such a point of view, to say the least.

The conclusions we have reached in our book are derived from known, experimentally and observationally tested properties of the Universe, including not least amongst them the property that living cells can replicate. The rival theory of the 'chemical evolution' of primitive life, and of the evolution of life to progressively higher levels entirely through random processes, is an uneasy combination of dogma and wishful thinking. To claim that there is evidence for chemical evolution in the fact that complex organic molecules have been discovered in meteorites or in planetary atmospheres is certainly wishful thinking. The most likely and well-tested process that could have led to meteoritic amino acids and lipids being formed must surely be biological processes operating in the body from which the meteorite was derived.

There have been many attempts at dismissing cosmic biology and its wider implications by one-line disproofs. All these attempts have inevitably ended in failure, often concealing efforts to substitute logic with polemics. A few critics have stated that the probability argument should not be taken seriously for the reason that exceedingly improbable events and arrangements do actually occur in nature. Take for instance the arrangement of molecules defining an intricate pattern within a cumulus cloud on a summer's day at a particular place and time. This precise arrangement, one could argue, is like the correct arrangement of amino acids in the enzymes, which is to say, exceedingly improbable. But such an argument misses the crucial point that, whereas enzymes have remarkable and even amazing chemical properties, there is nothing special about a particular arrangement of molecules in a cloud. One arrangment is no more significant than another. Consider another, more precisely calculable event of a similar type. Every Saturday afternoon during the rugby season pretty well the same set of rugby fans queue up to gain entry into the football stadium at Cardiff Arms Park. The fans would number several thousands, and the precise order of their entry through the gates must surely be a more or less random affair. Randomness would show as highly variable patterns of entry in the queue from one Saturday to the next. The number of different arrangements of fans entering could be shown to number $10^{40,000}$ if the total number of fans is somewhat larger than 10,000. It has been argued that such events do occur, and so the production of the enzymes is no big deal. But, again, the point being missed is that the unique arrangement of amino acids in the enzymes is special, whereas there is no sensible criterion that can distinguish between one Saturday's order of rugby fans and the next. If, however, we were to prescribe beforehand a list of fans in a particular order as being preferred or desirable for any reason, then we would, on the average, need to wait $10^{40,000}$ Saturdays before this order was realised by chance. Of course we could contrive to force the desired order of entry through the gate by persuasion or compulsion, but then intelligence would be involved.

Another line of defence adopted by the Darwinists has been to say that evolution is seen to take place in the present day without the need for an external input. On careful examination, each such claim is found to be based either on erroneous argument or incomplete knowledge of the detailed processes that are involved. For instance, the development of penicillin-resistant strains of bacteria has been cited. Yet a phenomenon of this kind need involve only a simple sieving out from a wide range of variants of a bacterial type that is present to start with. As we have mentioned in chapter 7, every space-incident type of bacterium or virus must have a definite genetic spread about some average genetic make-up. Penicillin therefore must act as a sieve to bring out and amplify a component from within a bacterial population that already contained a penicillin-resistant property.

Similar arguments are relevant in other areas of biology as well. It has been argued that the influenza virus evolves during the time intervals between the appearance of major antigenic shifts, in a way that indicates evolution in response to host pressure. It is clear from the way that influenza affects several species during epidemics that a wide genetic spread must exist within any viral reservoir that is supplied from comets. Copying errors generated as a virus is transcribed can be important in some cases, but even in those cases it is likely in our view that mutation within host cells is less important than the sieving out process that takes place as immunity develops in an animal population in a progressive, sequential manner. One would expect a more or less continual sorting out in response to immunity of particular genetic types from the original spread that existed in the cometary inocculant. Genetic trees constructed in an attempt to connect one viral type with another are in such a situation not meaningful in evolutionary terms.

The development of industrial melanism in moths is also cited as definitive evidence for internally generated evolution. Moths in pre-industrial conditions were pale coloured, but with the rise of pollution in industrial cities it was found that the moth population turned black over a few generations in a rather dramatic way. Here again an effect other than internally driven creative evolution is involved. There is no conceivable way in which the gene to produce the pigment melanin (the cause of the brown colour in our eyes, skin and hair) could be internally generated if the information for it, or a great part of it, was not already present. What is involved in this instance is that the gene for melanism existed in the pre-industrial moth but had been corrupted at just one or two genetic sites. The gene was therefore at the margin of genetic significance, temporarily inoperable but recoverable in a working condition through unfaithful copying at the one or two sites by which the gene had previously become inoperable. In other words, unfaithful copying from generation to generation happened by chance to turn a small fraction of the moths dark again, a condition favourable for survival against predation in industrial conditions.

Natural selection then caused the moth population to turn black over a number of generations. The essential point is that the recovery of melanin production was only possible because the gene in question was already at the edge of significance in the pale moth. If nothing related to melanin production had been present in the pale moth, the population would not have turned dark, because the relevant gene could not have been found *ab initio*.

The alternative to assembly of life by random, mindless processes is assembly through the intervention of some type of cosmic intelligence. Such a concept would be rejected out of hand by most scientists, although there is no rational argument for such a rejection. With our present knowledge, human chemists and biochemists could now perform what even ten years ago would have been thought impossible feats of genetic engineering. They could, for instance, splice bits of genes from one system to another, and work out, albeit in a limited way, the consequences of such splicings. It would not need too great a measure of extrapolation, or too great a licence of imagination, to say that a cosmic intelligence that emerged naturally in the Universe may have designed and worked out all the logical consequences of our own living system. It is human arrogance and human arrogance alone that denies this logical possibility.

To suppose that a life-form based on exactly the same basic system as ours, the same complex molecular jigsaw bits, had any part in this grander scheme of things would be to beg the question of origins again. The ultimate cosmic intelligence would need to be comprised of different units from those of our own life-form, possibly also units that are intrinsically more robust than ours, with an ability perhaps to withstand much higher temperatures. The essential complexity of our own cells and of the omnipresent cosmic bacteria must be due in part to the necessity to replicate. The bare essentials for intelligence and consciousness might be separable from such fragile structures, and the ultimate cosmic intelligence built from these more robust structures could well be thought to persist for exceedingly long timescales, even for an eternity. A prime requirement is that such an intelligence be capable of computation, analysis and exploration of the Universe at large. In a word it has to approximate to a condition of Omniscience. Our own intelligence might be thought to be limited, ultimately, by the number of intercommunicating nerve cells that are available, and the speed of communication between them. A cosmic intelligence could be envisioned on a much more ambitious scale than the capacity of our own brains.

These considerations are admittedly vague in their detail, but such logical constructs as they generally imply cannot be denied as possible solutions of this problem. The overt rejection of logic in relation to such matters, as is apparent in the present day, can be traced back to a series of historical accidents that happened over a century ago. In the middle years of the nineteenth century the Church had become a formidable social force to be reckoned with throughout most of Western Europe. The power of the Church provoked resentment in some circles, and the only way forward to become freed of what seemed to be its repressive regime was to attack the very foundation of its beliefs. To such an end an intellectual movement was launched that culminated in the publication of Darwin's *Origin of Species*. This book has been widely acclaimed and interpreted as being a justification for abandoning the biblical ideas of creation in favour of random processes. Such processes are thought initially to operate on inorganic chemicals leading to primitive life, and thereafter on living systems themselves to produce the spectacle of life in its entirety.

This socioscientific movement of the nineteenth century would not, we think, have gained ground but for the insistence on the side of the Church, and in particular of a few fundamentalist groups within it, that evolution itself had to be denied. To deny the fact that life evolved in a progressive and connected way from single-celled organisms to multicelled organisms to higher plants and animals, despite the well-proven facts of geology, was an invitation to disaster. Yet, even today, Christian fundamentalists insist that the biblical story of Genesis has to be regarded as literal truth, with no room for deviation, and that any science that goes against it is either deceptive or false.

The fear that Christian fundamentalism will raise its head to the detriment of science is perhaps a major reason for rejecting ideas such as are contained in the present book concerning the cosmic origins of life and intelligence. This is particularly true in the United States of America, where fundamentalists in the South are waging a battle with the Federal authorities to defend the teaching of the biblical story alongside the Darwinian explanation of evolution. Any weakening of the carefully erected Darwinian edifice, it is thought, would open the flood gates to fundamentalist dogma. If Darwinism was proven fact and all the fundamentalist dogma was proven falsehood, one might ask whether there would be any good reason for the paranoia that prevails. The truth must be that there is a lot that is basically wrong with Darwinism and a good deal that is in essence, though not in detail, right with the fundamentalist point of view.

Whatever the historical circumstances might be, there can be no justification at all for rejecting outright the concept of cosmic life and the logic of a creation. The facts clearly point in this direction. A society that so stubbornly fails to distinguish scientific truth from falsehood must inevitably be following a path to self-destruction, in our opinion.

Over the years the Darwinian thesis has been elaborated upon by successive generations of biologists, each generation deeply conscious of the need to maintain as rigid a stance as possible against the idea of a creation. Alfred Russel Wallace was a notable exception. Wallace, who had done perhaps more than anyone else to support the general concept of evolution by natural selection, came eventually to the conclusion that, while evolution through internally generated changes acted upon by selection works for some of the properties of plants and animals, for other properties it does not work. Wallace wrote thus:

> The law of Natural Selection or the survival of the fittest
> is, as the name implies, a rigid law, which acts by the life
> or death of the individuals submitted to its action. From its
> very nature it can act only on useful or hurtful
> characteristics, eliminating the latter and keeping up the
> former to a fairly general level of efficiency. Hence it
> necessarily follows that the characters developed by its
> means will be present in all the individuals of a species, and,
> though varying, will not vary very widely from a common
> standard .... In accordance with this law we find that all

those characters in man which were certainly essential to
him during his early stages of development, exist in all
savages with some approach to equality. In the speed of
running, in bodily strength, in skill with weapons, in
acuteness of vision, or in power of following a trail, all are
fairly proficient, and the differences of endowment do not
probably exceed the limits of variation in animals .... So, in
animal instinct or intelligence, we find the same general level
of development. Every wren makes a fairly good nest like its
fellows; even a fox has an average amount of the sagacity of
its race; while all the higher birds and mammals have the
necessary affections and instincts needful for the protection
and bringing up of their offspring.

But in those specifically developed faculties of civilised
man which we have been considering, the case is very
different. They exist only in a small proportion of individuals,
while the difference of capacity between these favoured
individuals and the average of mankind is enormous. Taking
first the mathematical faculty, probably fewer than one in
a hundred really possess it, the great bulk of the population
having no natural ability for the study, or feeling the slightest
interest in it.[1] And if we attempt to measure the amount
of variation in the faculty itself between a first-class
mathematician and the ordinary run of people who find any
kind of calculation confusing and altogether devoid of
interest, it is probable the former could not be estimated at
less than a hundred times the latter, and perhaps a thousand
times would more nearly measure the difference between
them.

The artistic faculty appears to agree pretty closely with
the mathematical in its frequency. The boys and girls, who
going beyond the mere conventional designs of children, draw
what they see, not what they know to be the shape of things;

---

[1] This is the estimate furnished to me by two mathematical masters in one of our
great public schools of the proportion of boys who have any special taste or capacity
for mathematical studies. Many more, of course, can be drilled into a fair knowledge
of elementary mathematics, but only this small proportion possess the natural faculty
which renders it possible for them ever to rank high as mathematicians, to take any
pleasure in it, or to do any original mathematical work.

who naturally sketch in perspective, because it is thus they
see objects; who see, and represent in their sketches, the light
and shade as well as the mere outlines of objects; and who
can draw recognisable sketches of every one they know, are
certainly very few compared with those who are totally in-
capable of anything of the kind. From some inquiries I have made
in schools, and from my own observation, I believe that those
who are endowed with this natural artistic talent do not exceed,
even if they come up to one percent of the whole population.

The variations in the amount of artistic faculty are
certainly very great, even if we do not take the extremes. The
gradations of power between the ordinary man or woman
'who does not draw', and whose attempts at representing any
object, animate or inanimate, would be laughable, and the
average good artist who, with a few bold strokes, can produce
a recognisable and even effective sketch of a landscape, a
street, or an animal, are very numerous; and we can hardly
measure the difference between them at less than fifty or a
hundred fold.

The musical faculty is undoubtedly, in its lower forms, less
uncommon than either of the preceding but it still differs
essentially from the necessary or useful faculties in that it
is almost entirely wanting in one-half even of civilised men.
For every person who draws, as it were instinctively, there
are probably five to ten who sing or play without having been
taught and from mere innate love and perception of melody
and harmony.[1] On the other hand, there are probably about
as many who seem absolutely deficient in musical perception,
who take little pleasure in it, who cannot perceive discords
or remember tunes, and who could not learn to sing or play
with any amount of study. The gradations, too, are here quite
as great as in mathematics or pictorial art, and the special
faculty of the great musical composer must be reckoned many
hundreds or perhaps thousands of times greater than that of
the ordinary 'unmusical' person above referred to.

It appears then, that, both on account of the limited number

---

[1] I am informed, however, by a music master in a large school that only about one
percent have real or decided musical talent, corresponding curiously with the estimate
of the mathematicians.

of persons gifted with the mathematical, the artistic, or the musical faculty, as well as from the enormous variations in its development, these mental powers differ widely from those which are essential to man, and are, for the most part, common to him and the lower animals, and that they could not, therefore, have been developed in him by means of the law of natural selection.

In addition to the clearly 'non-Darwinian' gifts referred to in the above quotation, as far as one can tell, humans are the only creatures on our planet endowed with a faculty that permits the formulation of abstract concepts removed from practical applications connected with survival. It is this faculty that led eventually to the emergence of 'technological' Man. Humans are also the only creatures endowed with what could be called a religious instinct. We assign an importance to ourselves beyond the immediate dictates of survival. We think of our kith and kin and all fellow members of our species as having an importance as individuals, as well as an importance that transcends individual identities. We have an innate yearning, it seems, to be identified as a part of some ill-defined grander scheme of things. We are purposive creatures, and to discover an ultimate purpose that links us one to another and to the wider Universe, Man has traditionally turned to religion.

Religion is probably as old as Man himself. Every religion that has evolved can be seen as an effort to discover a greater purpose for which we might live and die, and to seek a First Cause for all living things. In many ways these basic religious aspirations are convergent with our own more rational endeavours as scientists to understand the Universe and the Origin of Life within it. Interestingly, the result we have now arrived at, namely the logical need for intelligence in the universe, is also consistent with the tenets of most of the major religions of the world.

Such an idea was already considered in philosophical terms by Aristotle (384–322 BC) as part of his *Metaphysics*. Aristotle introduced the concept of Prime Mover or God, although this was prompted in part by his incomplete knowledge of physics, in his search for first causes to maintain such phenomena as the motions of planets. Aristotle's philosophical writings when they were re-discovered at the beginning of the thirteenth century were considered to be such a threat to narrowly interpreted Christian beliefs that they were banned by various ecclesiastical authorities for over fifty years. Later Thomas Aquinas, in his work *Summa Theologica*, built upon the philosophical foundations of Aristotle to argue that the existence of God could be 'proved' within the Aristotelean framework.

The Creator has been given many shapes and names in the diverse cultures throughout the world. He has been called Jehovah, Brahma, Allah, Father in Heaven, God, in different religions, but the underlying concept has been the same. The general belief that is common to all religions is that the Universe, particularly the world of life, was created by a 'being' of incomprehensibly magnified human-type intelligence. It would be fair to say that the overwhelming majority of humans who have ever lived on this planet would have instinctly accepted this point of view in some form, totally and without reservation. In view of the thesis of this book, it would seem to be almost in the nature of our genes to be able to evolve a consciousness of precisely this kind, almost as if we are creatures destined to perceive the truth relating to our origins in an instinctive way.

Bibliography

AbadiH. and Wickramasinghe,N.C., Pre-
biotic Molecules in Martian Dust
Clouds, *Nature, 267*, 687 (1977)

Anderson,A.W., Nordan,H.C., Cain,R.F.,
Parish,G. and Duggan,D., Studies on a
Radio-Resistant *Micrococcus*. 1.
Isolation, Morphology, Cultural
Characteristics and Resistance to
Gamma Radiation, *Food Technology,
10*, 575 (1956)

Arrhenius, Svante, The Propagation of
Life in Space, *Die Umschau, 7* (1903)

Bernal,J.D., *The Origin of Life* (World
Publishing Co., Cleveland, 1967)

Bowen,E.G., An Unorthodox View of the
Weather, *Nature, 177*, 1121 (1956)

Cocconi,G. and Morrison,P., Searching for
Interstellar Communications, *Nature,
184*, 844 (1959)

Crick, Francis, *Life Itself* (Macdonald &
Co., London, 1981)

Darwin, Charles, *The Origin of Species*
(John Murray, London, 1872)

DrakeF.D., Project Ozma, *Physics Today,
14*, 40 (1961)

Dyson,F.J., Search for Artificial Stellar
Sources of Infrared Radiation, *Science,
131*, 1667 (1960)

Gold, Thomas, *Power from the Earth*
(J.M.Dent & Sons, London, 1988)

Goldsmith, Donald, *The Quest for
Extraterrestrial Life: A Book of Readings*
(University Science Books, Mill Valley,
Calif., 1980)

Greeley,R., *Planetary Landscapes* (Allen &
Unwin, London, 1985)

Hart,M.H., An Explanation for the
Absence of Extraterrestrials on Earth,
Quarterly Journal Roy. Astr. Soc., 16,
128 (1975)

Heinrich,M.R., (ed). *Extreme
Environments: Mechanisms of Microbial
Adaptation* (Academic Press, 1976)

Hoover,R.B., Hoyle,F.,
Wickramasinghe,N.C., Hoover,M.J. and
Al-Mufti,S., Diatoms on Earth, Comets,
Europa and in Interstellar Space, *Earth,
Moon and Planets, 35*, 19 (1986)

Horowitz,N.H., The Search for Life on
Mars, *Scientific American, 237*, 52
(1977)

Hotckin,J., Lorenz,P. and Hemmenway,C.,
Survival of Micro-organisms in Space,
Nature, 206, 442 (1965)

Hoyle, Fred, and Narlikar, Jayant, *The
Physics-Astronomy Frontier* (W. H.
Freeman Co., San Fransisco, 1980)

Hoyle, Fred, Wickramasinghe, Chandra
and Watkins, John, *Viruses From Space*
(University College Cardiff Press, 1986)

Hoyle,F., Wickramasinghe,N.C. and
Pflug,H.D., An Object within a Particle
of Extraterrestrial Origin compared
with an Object of Presumed Terrestrial
Origin, *Astrophys. Sp. Sci., 113*, 209
(1985)

Hoyle,F., Wickramasinghe,N.C., Al-
Mufti,S. and Olavesen,A.H., Infrared
Spectroscopy of Micro-organisms near
3.4 microns in relation to Geology and
Astronomy, *Astrophys. Sp. Sci., 81*, 489
(1982)

Hoyle,F., Wickramasinghe,N.C. and Al-Mufti,S., Organo-siliceous Biomolecules and the Infrared Spectrum of the Trapezium Nebula, *Astrophys. Spc. Sci.*, *86*, 63 (1982)

Hoyle,F., Wickramasinghe,N.C. and Al-Mufti,S., The Viability with respect to Temperature of Dry Micro-organisms incident on the Earth's Atmosphere, *Earth, Moon and Planets*, *35*, 79 (1986)

Hoyle,F., Wickramasinghe,N.C. and Wallis,M.K., On the Nature of Dust Grains in the Comae of Comets Cernis and Bowell, *Earth, Moon and Planets*, *33*, 179 (1985)

Hoyle,F., Wickramasinghe,N.C. and Watkins,J., Legionnaires' Disease: Seeking a Wider Cause, *The Lancet*, 25 May 1985, p. 1216

Hoyle,F. and Wickramasinghe,N.C., Dust in Supernova Explosions, *Nature*, *226*, 62 (1970)

Hoyle,F. and Wickramasinghe,N.C., Primitive Grain Clumps and Organic Compounds in Carbonaceous Chondrites, *Nature*, *264*, 45 (1976)

Hoyle,F. and Wickramasinghe,N.C., Polysaccharides and the Infrared Spectra of Galactic Sources, *Nature*, *268*, 610 (1977)

Hoyle, Fred and Wickramasinghe, Chandra, *Lifecloud* (J.M.Dent & Sons, London, 1978)

Hoyle,F. and Wickramasinghe,N.C., On the Nature of Interstellar Grains, *Astrophys. Sp. Sci.*, *66*, 77 (1979)

Hoyle, Fred and Wickramasinghe, Chandra, *Diseases from Space* (J. M. Dent & Sons, London, 1979)

Hoyle, Fred and Wickramasinghe, Chandra, *Evolution from Space* (J.M.Dent & Sons, London, 1981)

Hoyle, Fred and Wickramasinghe, Chandra, *Space Travellers: the Bringers of Life* (University College Cardiff Press, 1981)

Hoyle,F. and Wickramasinghe,N.C., Comets – a Vehicle for Panspermia, in *Comets and the Origin of Life*, ed. C. Ponnamperuma (D.Reidel Publishing Co., 1981)

Hoyle,F. and Wickramasinghe,C. Why Neo-Darwinism does not Work (University College Cardiff Press, 1982)

Hoyle,F. and Wickramasinghe,N.C., A Model for Interstellar Extinction, *Astrophys. Sp. Sci.*, *86*, 321 (1982)

Hoyle,F. and Wickramasinghe,N.C., Proofs that Life is Cosmic, *Mem. Inst. Fund. Studies*, Sri Lanka, No. 1 (1983)

Hoyle, Fred and Wickramasinghe, Chandra, *From Grains to Bacteria* (University College Cardiff Press, 1984)

Hoyle, Fred and Wickramasinghe, Chandra, *Living Comets* (University College Cardiff Press, 1985)

Hoyle, F. and Wickramasinghe,N.C., Some Predictions on the Nature of Comet Halley, *Earth, Moon and Planets*, *36*, 289 (1986)

Hoyle, Fred and Wickramasinghe, Chandra, *Archaeopteryx, the Primordial Bird* (Christopher Davies, 1986)

Hoyle,F. and Wickramasinghe,N.C., The Case for Life as a Cosmic Phenomenon, *Nature*, *322*, 509 (1986)

Hoyle, F. and Wickramasinghe,N.C., On the Nature of Interstellar Grains, *Q.Jl.R.A.S.*, *27*, 21 (1986)

Hoyle,F. and Wickramasinghe,N.C., Organic Dust in Comet Halley, *Nature*, *328*, 117 (1987)

Hoyle,F. and Wickramasinghe,N.C., Cometary Organics, *Nature*, *331*, pp. 123 and 666 (1988)

Hoyle,F. and Wickramasinghe, N.C., Interstellar Extinction by Cometary Organic Grain Clumps, *Astrophys. Space Sci.*, *140*, 191 (1988)

Hoyle,F. and Wickramasinghe,N.C., Metallic Particles in Astronomy, *Astrophys. Sp. Sci.*, *142* (1988)

Hoyle,F. and Wickramasinghe,N.C., Interstellar Extinction by Cometary Organic Grain Clumps, *Astrophys. Sp. Sci.*, *140*, 191 (1988)

Hoyle, Fred, *Astronomy and Cosmology* (W.H.Freeman & Co., San Francisco, 1975)

Hoyle, Fred, *Energy or Extinction?* (Heinemann, London, 1977)

Hoyle, Fred, *The Cosmogony of the Solar System* (University College Cardiff Press, 1978)

Hoyle, Fred, *The Intelligent Universe* (Michael Joseph, 1983)

Jackson, Francis and Moore, Patrick, *Life in the Universe* (Routledge & Kegan Paul, London, 1987)

Jeans, Sir James, Is There Life on the Other Worlds? *Science*, *95*, 589 (1942)

Levin,G.V. and Straat,P.A., A Search for a Nonbiological Explanation of the Viking Labeled Release Life Detection Experiment, *Icarus*, 45, 494 (1981)

Lowell, Percival, *Mars as the Abode of Life* (Macmillan Co. NY, 1908)

Mendis,D.A., and Wickramasinghe,N.C., The Composition of Cometary Dust: The Case Against Silicates, *Astrophys. Sp. Sci.*, 37, L13 (1975)

Mitchell,F.J. and Ellis,W.L., Surveyor III: Bacterium Isolated from Lunar-Retrieved TV Camera, *Proc. Second Lunar Science Conf.*, 3, 2721 (1971)

Moseley,B.E.B., Repair of Ultraviolet Radiation Damage in Sensitive Mutants of *Micrococcus radiodurans*, *Journal of Bacteriology*, 97 (1967)

Nagy,B., Fredriksson,K., Urey,H.C., Claus,G., Anderson,C.A. and Percy,J., Electron Probe Microanalysis of Organised Elements in the orgueil Meteorite, *Nature*, 198, 121 (1963)

Narlikar,J.V. and Wickramasinghe,N.C., Microwave Background in a Steady-state Universe, *Nature*, 216, 43 (1968)

Narlikar,J.V. and Wickramasinghe,N.C., Interpretation of the Cosmic Microwave Background, *Nature*, 217, 1235 (1968)

Papagainnis,M., Are We Alone, or Could They be in the Asteroid Belt? *Quarterly J. of R. Astr. Soc.*, 19, 277 (1978)

Papagaiannis,M.D., Recent Progress and Future Plans on the Search for Extraterrestrial Intelligence, *Nature*, 318, 135 (1985)

Pflug,H.D., Ultrafine Structure of the Organic Matter in Meteorites, in C. Wickramasinghe ed., *Fundamental Studies and the Future of Science* (University College Cardiff Press, 1984)

Spencer Jones, Sir Harold, *Life on Other Worlds* (Hodder & Stoughton, 1940)

Urey,H.C., Life-forms in Meteorites, *Nature*, 193, 119 (1962)

Vanysek,V. and Wickramasinghe,N.C., Formaldehyde Polymers in Comets, *Astrophys. Sp. Sci.*, 33, L19 (1975)

Wickramasinghe,N.C., Edmunds,M.G., Chitre,S., Narlikar,J.V. and Ramadurai,S., A Dust Model for the Cosmic Microwave Background, *Astrophys. Sp. Sci.*, 35, L9 (1975)

Wickramasinghe,N.C., Wallis,M.K., Al-Mufti,S., Hoyle,F., and Wickramasinghe,D.T., The Organic Nature of Cometary Grains, *Earth, Moon and Planets*, 40, 101 (1988)

Wickramasinghe,N.C., *Interstellar Grains* (Chapman & Hall, Lond. 1967)

Wickramasinghe,N.C., Formaldehyde Polymers in Interstellar Space, *Nature*, 252, 462 (1974)

Wickramasinghe, Chandra, *The Cosmic Laboratory* (University College Cardiff Inaugural Lecture, 1975)

Wickramasinghe, Chandra, *Is Life an Astronomical Phenomenon?* (University College Cardiff Press, 1982)

Wickramasinghe, Chandra, (ed) *Fundamental Studies and the Future of Science* (University College Cardiff Press, 1984)

Index